# Facility Manager's Portable Handbook

# Facility Manager's Portable Handbook



Bernard T. Lewis

McGraw-Hill

New York San Francisco Washington, D.C. Auckland Bogotá
Caracas Lisbon London Madrid Mexico City Milan
Montreal New Delhi San Juan Singapore
Sydney Tokyo Toronto

**Library of Congress Cataloging-in-Publication Data**

Lewis, Bernard T.
Facility manager's portable handbook / Bernard T. Lewis.
p. cm.
ISBN 0-07-135121-3
1. Facility management—Handbooks, manuals, etc. 2. Plant engineering—Handbooks, manuals, etc.
TS184.L49 1999
658.2—dc21 99-33971
CIP

*McGraw-Hill*

*A Division of The* ***McGraw-Hill*** *Companies*



1 2 3 4 5 6 7 8 9 0 DOC/DOC 9 0 4 3 2 1 0 9

ISBN 0-07-135121-3

*The sponsoring editor for this book was Linda Ludewig, the editing supervisor was Suzanne Ingrao, and the production supervisor was Pamela A. Pelton. It was set in Century Schoolbook per the MHT 5 x 8 design by Joanne Morbit and Paul Scozzari of McGraw-Hill's Hightstown, N.J., Professional Book Group composition unit.*

*Printed and bound by R. R. Donnelley & Sons Company.*

This book is printed on recycled, acid-free paper containing a minimum of 50% recycled de-inked fiber.

*In presenting the* Facility Manager's Portable Handbook, *I wish every success to facility managers, and all others, who, day-in and day-out, engage in essential facility engineering, operations, and maintenance functions. Without their efforts, this country would not, and could not, remain competitive.*

# Contents

# Preface

*BUILDINGS SHOULD BE SEEN AND NOT HEARD. Buildings talk. They send us messages through unexpected repairs, equipment breakdowns, tenant complaints, system failures, high operating expenses, staff turn-over, and many other problems that can trouble today's facility managers.*

The *Facility Manager's Portable Handbook,* based upon proven concepts of good facilities operations and maintenance management and engineering principles, procedures, systems, and techniques, provides the keys to successful building operations. This handbook presents the subjects of facilities operations and maintenance functions within a framework that can provide a practical approach to the solving of problems of a facility department's organizational design, and systems and procedures operational problems. It also serves as a guide to the specific problems of methodology and direction of the facilities operations and maintenance functions themselves.

The ideas presented in this handbook can be immediately useful to any facility manager in analyzing those organizational (functional) areas for which he or she is now, or may ultimately become, responsible. It should aid him or her in interpreting problems and relating them to their sources and causes. Most of all, it can serve as a signpost pointing to the general areas of proven management thinking that help control the facilities operations and maintenance functions costs and effectiveness.

Operations and maintenance functions are the most important portions of a good facilities computerized maintenance management system (CMMS). They can be the bridge that spans the division between the organization's mission accomplishment and facilities department operations. These facilities' functions are much more than periodic inspections of equipment, systems, and structures to avoid breakdowns, and the rapid response to emergency service calls (normally called fire fighting), although these are essential portions of any good facilities management program.

As industry turns more and more toward automation and robotics used in processing equipment in which there is a very high capital investment, it is of the utmost importance that maximum utilization of the facility's equipment, systems, and structures be attained, and at the least overall cost. The main areas for facilities management consideration are overall equipment and systems utilization, maintenance, and repair costs. To attain maximum production effort, there must be a minimum of downtime due to faulty equipment, systems, or structures. The cost of this type of production "insurance" can only be evaluated when it is weighed against the costs of probable and expected production delays due to machinery, equipment, and/or systems breakdowns. Often the instinctive reactions of facility managers will be to apply good engineering principles and strive only for good operations and maintenance performance. *But the real key to a cost-effective and profitable facility operations and maintenance program is to consider economics first and engineering second.* After a careful economic analysis, it will be seen that the good facility operations and maintenance program often pays its way by the maintenance it avoids and the reductions in operations costs it achieves.

In order to effect the "prevention and predictive" phases of the facility operations and maintenance program, there must be preplanned and prescheduled inspection, maintenance, repair, and overhauls for all equipment and systems. This also applies to the efficient, economical, and reliable operation of all mechanical, electrical, plumbing, utility, and fire safety equipment and systems. The purpose is efficient trouble-shooting to prevent major breakdowns that can cost an organization large sums of money. This planning and scheduling takes into account the effect of preventive and predictive maintenance inspections on the regular maintenance schedule as well as the organization's production schedule. Both the maintenance and mission accomplishment schedules can be disrupted by inspections and overhauls. *The objective is to limit the inspection and overhaul cycle times, to avoid disrupting an organization's mission accomplishment, to the minimum time allowed according to the cognizant standard data time allowances.*

This handbook is concerned with the functions and procedures related to outsourcing; operations and maintenance plans; preventive and predictive maintenance procedures; operations and maintenance procedures; custodial services; landscaping services; elevator and escalator services; water treatment services; and architectural, structural, and sustaining maintenance and repair services. This handbook presents examples of currently used proven systems, techniques, and procedures that can be applied to the installation and effective utilization of a good operations and maintenance management program in any organization's facility department.

To keep this handbook portable, the treatment of some subjects has been necessarily brief, but subjects of greater importance have been dealt with more fully. The various tables and figures have been selected with great care, and only those that are most likely to be consulted daily by the facility manager have been included.

Care has been taken to arrange the chapters in a convenient and logical manner. A complete index is provided to further increase the facility with which any subject may be located.

Regardless of the property—residential, commercial, industrial, institutional, or government—systems, procedures, tables, and figures can be found in this concise handbook to enable the facility manager's daily work to go more smoothly. Furthermore, solutions to everyday facility management, engineering, systems, and equipment problems may be found practically at a glance, thus reducing normal problem-solving times.

*Facility managers cannot afford to be without this portable handbook.*

*Bernard T. Lewis*

# How to Use This Handbook

The concept of this book is that of a *personal tool* that summarizes, in a readily available manner, the 20% of the information and data needed 80% of the time by *facility managers* in the course of pursuing their day-to-day activities of managing and controlling the facility's operations and maintenance tasks.

This handbook is meant to be at the facility manager's *fingertips* (open on his or her desk, carried in a briefcase, or carried in a pocket). It is not meant to sit on a bookshelf. It is meant to be used *every day* as a quick adjunct to assist in solving confronted existing or potential problems of any type—facilities engineering, operations, maintenance, etc.

This handbook is designed with both checklists and optimal functional systems and procedures descriptions format and to be quickly read, evaluated, and measured against operations and maintenance problems at hand, as well as potential problems. Detailed checklists are presented in Chaps. 1 and 11. Chapters 2 through 10 cover operations and maintenance functions in sufficient detail to provide solutions to problems discovered by using the checklists in Chaps. 1 and 11. Thus the facility manager will be able to measure and improve the following critical departmental factors:

- Effectiveness of management and fiscal control
- Use of sound organizational principles
- Policies and practices
- Systems and procedures

- Operations effectiveness
- Effectiveness of personnel policies and practices

This book *is not a substitute* for professional expertise or other more detailed and specialized books. It will serve as a continuing everyday aid that presents the *highlights* of other available sources.

# Facility Manager's Portable Handbook

Chapter

# 1

# Overall Program Performance Review

## Introduction

Facility managers are being forced to find new ways and means of reducing their operational costs due to increased pressure being exerted by top management for organization-wide cost reductions. There are only two avenues open to the facility manager in this respect: (1) squeeze more operations and maintenance effort from his or her current work force, or (2) obtain the same operations and maintenance effort from a reduced work force.

As in all decision-making actions, the facility manager must first identify the problem, and then break it into its smallest definable components for detailed corrective actions. To the facility manager this means asking the question, "What must be done to introduce management control systems and procedures in my department with a minimum of disruption—keeping in mind union problems and the usual subordinate supervisor's feeling of apathy and distrust toward the introduction of any new management control systems and procedures?"

On the surface, facility departments in most organizations appear to function properly until an organization-wide cost-reduction program discloses excessive operations and maintenance costs and customer service complaints. Where to look and what to do for *corrective action* will be covered in the following sections. Covered will be facility department strategic objectives, operations and maintenance performance audit criteria, comparative statistics of facility department operations, and departmental management audit criteria. *This approach is key to a successful operations and maintenance cost reduction and productivity effectiveness improvement program.*

## Strategic Objectives Planning

To be a good facility manager one must be able to *plan, organize, direct,* and *control,* and one must be able to deal with *people.* Buildings, machinery, equipment, and systems need little or no direction or control, but people do. People want and need help in obtaining objectives. Thus the term "strategic objectives planning" refers to people and dollars. *Planning* involves formulation and looking ahead. *Strategic objectives planning* answers the questions: What is to be done? How? When and who? Strategic objectives planning is a preliminary activity where the facility manager can formulate long range ideas, goals, and objectives. It is the first step a facility manager must take to change the department's culture and management style. Without good strategic objectives planning, most departmental objectives will never be attained. *The key to success in this regard is to ensure that once strategic objectives are formulated and approved, they must receive widespread facility department dissemination and acceptance.*

Selected strategic objectives are shown in Fig. 1.1.

## Departmental Operations, Maintenance, and Management Audit Criteria

Determining and discovering methods, procedures, techniques, equipment, and systems problem (current and potential) areas, in the operations and maintenance functions, will be aided by careful analysis and evaluation of the following

1. Provide for and implement control inspections of all organization facilities (buildings, machinery, equipment, and systems) on a three-year scheduled basis. Control inspections are defined as complete and thorough inspections of all facilities from roof to ground on a predetermined schedule.
2. Transition from a majority reliance on in-house work-force resources to a better balance of contract–to–in-house work resources to maximize utilization of both resources. Target departmental functions, initially, that are routine and repetitive in work-force performance (i.e., custodial, landscaping, elevators and escalators, etc.).
3. Institute a computerized maintenance management system (CMMS) that will record all work requested and performed by the department as well as all relevant data, provide accurate workload and scheduling projections, provide accurate facilities conditions information and data, and provide personnel provide budgetary information.
4. Institute an energy management control system (EMCS) that will control and monitor heating, ventilating, and air conditioning (HVAC) use to enable cost-effective and cost-efficient operations without degradation of the organization's environmental requirements. Additionally, ensure that replacement equipment and systems items are of the most energy-efficient varieties.
5. Develop a roof management program to include a complete roof inventory and condition analysis, long-term replacement and recondition plans, and an annual preventive and predictive maintenance inspection plan.
6. Emphasize maintenance activity while systematically reducing alterations, improvements, remodeling, and other nonmaintenance activities. Simultaneously upgrade current facilities structural and HVAC conditions.
7. Increase effective response times to emergency/service work requests by ensuring better information transmittal, and initiating a special organizational unit that responds only to emergency/service work requests. This approach will reduce both the response times and the cost of performing these tasks.
8. Reduce customer work requests to the facility department from its current percentage level to less than 25 percent for maintenance and repair items. This will allow the department to go from a "fire fighting" to a "planned" method of operation for maintenance and repair items. Institution of control inspections as indicated in item 1 of this list will assist in meeting this objective.

**Figure 1.1** Selected strategic objectives.

Operations Management Audit Checklist and Maintenance Management Audit Checklist; Tables 1.1 through 1.9, Comparative Statistics of Facility Department Operations; and the chapter-ending Management Audit Checklist.

### Operations Management Audit Checklist

1. Does the Building Operations Plan describe standards and guidelines for operating building HVAC equipment such that all specified temperature, lighting, and other environmental conditions are provided during buildings operating hours (Monday through Friday except for holidays and organizational elements that have to operate 24 hours per day, 365 days per year)?
2. Are systems and equipment placed in the automatic mode to the extent practical and feasible?
3. Are routine procedures established and used to guide start-ups and shutdowns?
4. Are steps taken to ensure that equipment is not cycled at frequencies that will lead to premature failure?
5. Are steps taken to ensure that when duplicate systems exist, equipment is rotated to equalize usage?
6. Are utility meters read on the same day as they are read by the power company?
7. Are inventories maintained for all fuels received, stored, and used?
8. Are facility start-up procedures varied according to external environmental conditions and average office space temperatures?
9. Are the air-conditioning and heating equipment started (and shutdown) at the same time each day?
10. Does the start-up Operating Engineer assess external environmental conditions (weather, light, etc.), and any other cognizant building-use factors, to ensure that the energy management control system has started or will start the mechanical equipment in order to have the

building within prescribed environmental limits by the start of normal building operating hours?

11. Are chillers, and all auxiliary equipment, operated during the cooling season, and as needed throughout the year, in accordance with outside ambient temperatures, through the energy management control system, to maintain prescribed building(s) interior cooling temperatures?
12. Are air handling units operated 12 months of the year, between the hours of 7:00 A.M. and 5:00 P.M.? Are air handlers operated on the economizer cycle (utilizing outside air) when the outside ambient temperature is below 60°F, and in conjunction with the chillers as needed when the outside temperature exceeds 60°F? Is outside air used to the maximum extent possible during moderate temperature periods, base upon outside air temperature and relative humidity conditions?
13. After ensuring that the facility has reached prescribed environmental limits, are the Operating Engineers assigned additional operations and maintenance tasks?
14. Has the Engineering Supervisor been instructed that before he leaves for the day, he will review with the energy management control system Operator as to what equipment is to be shut down for the night, and when to begin the shutdown procedures; and, also, what equipment is to be started up in the morning, and what time to begin the start-up procedures?
15. Are equipment start-up and shutdown procedures, as well as operating schedules, established to match normal load variations occurring during the regular course of a prescribed period (day/week/month/season/year)?
16. Are frequently used equipment and critical systems started prior to the peak-load period, in order to prevent "start-up spikes" at the peak curve?
17. Are major equipment or systems, including elevators, escalators, security , and fire alarm systems, not operational by the start of normal building operating hours reported to the Facility Manager no later than 7:45 A.M.?

18. Have procedures been installed specifying that during prolonged shutdown periods (weekends and holidays) for both summer and winter start-ups, that it may be necessary to advance start-ups by 1 to 2 hours, based on conditions at the time?
19. Have steps been taken to assure that all mechanical, electrical, plumbing, and utility systems are operated in an energy efficient manner, to provide the required environmental conditions, consistent with current federal, state, and local requirements?
20. Are temperature controls set to maintain an indoor environment that is consistent with general industry standards; and provides generally acceptable temperatures established for human comfort?
21. During off-duty hours are temperature controls maintained at a level that ensures the protection of the facility and its systems?
22. Are steps taken to ensure that ventilation and conditioned air supply, to all occupied, and nonoccupied spaces, is adequately filtered during normal hours of occupancy to assure a safe and healthy environment?
23. Is outside air utilized, whenever possible (based on outside temperature and humidity levels), to maintain specified office and shop spaces temperature ranges? During moderate weather, outside air, mechanical economizers, and/or other energy saving equipment installed, should be used to the maximum extent possible.
24. Is building ventilation provided to the maximum extent allowed by the configuration and design of mechanical equipment installed in the buildings?
25. Is air adequately filtered at all times by using only air filters capable of at least 50 percent particulate removal? Are filters changed frequently, in accordance with best industry standards and practices?
26. Are environmental conditions maintained in special areas, such as photographic laboratories and computer rooms, to ensure the reliable operation of the program

equipment as recommended by the applicable equipment manufacturers?

27. Is domestic hot water maintained at 105°F?

28. Are lighting systems maintained to achieve the following illumination levels during normal organization work hours?

| | |
|---|---|
| Public areas within buildings: | 10 foot-candles |
| Normal workstations: | 50 foot-candles |
| General work areas: | 50 foot-candles |
| Storage areas: | 10 foot-candles |

Are lighting systems maintained facility-wide, at a level necessary for safety and security, at all times?

29. Is the energy management control system programmed to ensure that all unnecessary lighting is turned off, daily, throughout all buildings?

30. Are running test checks performed on large or high-energy-use equipment such as chillers, pumps, and air-handling equipment during normal core hours of operations? Are the running test checks prescheduled prior to performance to avoid an interruption in service, spikes in utility use, or increase in monthly electrical demand costs? Certain peak operational usage periods will be prohibited for running these tests or checks.

### Maintenance Management Audit Checklist

1. Do we have simple, clear-cut Work Order forms and work management and control procedures?
2. Are processed work requests classified as to maintenance, repair, alterations, or improvement?
3. Does the Work Request form have a “date wanted” space used for realistic dates and not for “at once,” “as soon as possible,” “rush,” etc.
4. Is each Work Order cost estimated in terms of man-hours by craft, and materials required, for each Work Center’s operations?

5. Are the materials, supplies, and parts required for the job listed, on a Bill of Materials, prior to starting work on each Work Order?
6. Are Critical Path Scheduling Techniques used for large Work Orders (those over 40 man-hours in duration)?
7. Are Work Center scheduling procedures used by craft and trade Foremen to lay out an advance days work for each craftsman each day?
8. Are progress charts maintained for large Work Orders during execution (those over 40 man-hours)?
9. Are Work Order completion dates promised and met?
10. Are equipment, systems, and machinery Preventive and Predictive Maintenance records maintained for all items?
11. Do the Preventive and Predictive Maintenance records show original and upkeep items for each item of equipment, systems, and machinery?
12. Are equipment, systems, and machinery maintenance and repair costs and data analyzed for the purpose of eliminating repetitive repairs?
13. Is all equipment, systems, and machinery inspected at fixed intervals?
14. Is lubrication work laid out on a route basis? Are oil changes "called out" systematically by the CMMS?
15. Is a Work Request initiated as soon as a repair is needed?
16. Is man-hour by craft backlog developed each week? Is this data reviewed weekly for necessary corrective actions as required?
17. Can we predict, in advance, the need for manpower additions or cutbacks? For overtime? For outsourcing?
18. Is the required number of craft man-hours reserved and scheduled for routine inspection tours?
19. Is the Storeroom's Work Order Materials Staging Area managed by an Inventory Control System as a module in the CMMS?

20. Have the Storeroom's Work Order Materials stock items been reviewed, during the last six months, and obsolete items been disposed of?
21. Have the various Work Centers' materials "Gold Piles" been reviewed during the past six months, and actions taken to update the items, including disposing of obsolete items? Has a Storeroom Stock catalogue been published showing an up-to-date inventory listing? Is it updated monthly?
22. Are the maintenance facilities, tools, and equipment good enough to do a first-class maintenance job?
23. Is time lost by craftsmen having to walk to and from storerooms to obtain required materials?
24. Do maintenance managers and supervisors meet at least once a week to discuss plans, progress, and problems?
25. Do maintenance supervisors report to the Work Control Manager actual conditions found in buildings, during maintenance work performance, that differ from conditions described on building plans? This information and data are required to aid the Facility Design Manager to keep "as-built drawings" up-to-date.
26. Do weekly Work Order reports compare actual work performance with planned work performance based on engineered performance standards?

## Comparative Statistics of Facility Department Operations

### Introduction

Good facility managers intuitively feel the need for performance indicators of their own functions, or areas of responsibility, as well as those of their subordinates. Since the control of performance must be preceded by measurement and comparison, most indicators are in the form of indexes. Most indexes are mathematical functions, usually ratios, of two or more quantitative or parametric measurements. Since data of this type are either not always routinely available or are of factors not amenable to direct measurement, the use of nonparametric determinations or estimates is sometimes necessary. A good example of a matter of management concern that is not directly measurable is the state of worker morale. Fortunately, indicators of this type are in the minority, and some type of direct or indirect measurement is usually possible.

The main problem confronting most facility managers is usually not a lack of data, but of a rationale to enable them to establish a basis and determine the best indicators for each of the factors they wish to keep under surveillance. A facility manager who has no basis for comparison or way to judge the adequacy of the management controls he or she has devised or inherited cannot adequately perform his or her function; that of providing optimum, cost effective operations and maintenance support to the facility. There is no easy cure for this axiom; however, the goal for providing relief in this respect is to *find a set of indexes that is adequate to represent all facets of the operation and to minimize distortion due to unbalanced emphasis on the various facets.*

The problem most frequently encountered is that facility managers are not making best use of the information that is currently being generated within the organization. This is particularly true in the facility department. In this instance, it occurs most frequently as a failure to use the information available at the total organization level or from adjacent function or departments such as production, operations, or whatever name is applied to the facility department's best customer and reason for existence.

### Steps in developing sets of indicators

In selecting a set of performance indicators for routine reporting, the facility manager should consider each index separately and, upon its adoption, check to see that a balanced set has been adopted for each sector of concern.

Steps involved in selecting a particular index include answering the questions:

1. What do I want to control?
2. What index will indicate the performance?

After selecting the index, a target value of desired performance is needed for comparison. These may be based on budgets, past history, future goals and objectives, or some other basis. Target values selected should be checked for consistency between indexes within the set.

Typical reports found to useful in industry, composed of groups of performance indicators for various sectors of facility management concern, are presented in Tables 1.1 through 1.9. *These indicators are not all inclusive. They are samples for each sector that can be used by the facility manager to expand coverage for his or her department.*

**TABLE 1.1 Work Order Control Sector**

| Number | Index | Units |
|---|---|---|
| 1 | $\dfrac{\text{Cumulative work orders completed}}{\text{Cumulative work orders received}}$ | Percent |
| 2 | $\dfrac{\text{Backlog work content, man-hours}}{40 \times \text{number of workers available}}$ | Number Weeks |

**TABLE 1.2 Backlog Control Sector**

| Number | Index | Units |
|---|---|---|
| 1 | $\dfrac{\text{Backlog work content, man-hours, skill “A,” etc.}}{40 \times \text{number of skill “A,” etc., workers available}}$ | Weeks |
| 2 | $\dfrac{\text{Active backlog work content, man-hours, skill “A,” etc.}}{8 \times \text{number of skill “A,” etc., workers available}}$ | Days |

**TABLE 1.3 Planning and Scheduling Control Sector**

| Number | Index | Units |
|---|---|---|
| 1 | $\frac{\text{Actual time charged to scheduled work, man-hours}}{\text{Scheduled time for scheduled work, man-hours}}$ | Percent |
| 2 | $\frac{\text{Total overtime worked, man-hours}}{\text{Total work performed}}$ | Percent |

**TABLE 1.4 Contractor Control Sector**

| Number | Index | Units |
|---|---|---|
| 1 | $\frac{\text{Number of contractor workers}}{\text{Total number of facility department workers}}$ | Percent |
| 2 | $\frac{\text{Contract man-hours worked by craft}}{\text{Total man-hours worked by craft}}$ | Percent |

**TABLE 1.5 Material Control Sector**

| Number | Index | Units |
|---|---|---|
| 1 | $\frac{\text{Requisitions filled from stock without delay}}{\text{Total requisitions received}}$ | Percent |
| 2 | $\frac{\text{Value of materials issued}}{\text{Value of materials inventory}}$ | Percent |

**TABLE 1.6 Organization Control Sector**

| Number | Index | Units |
|---|---|---|
| 1 | $\frac{\text{Average number of facility department workers}}{\text{Facility value (replacement cost)}}$ | Number per \$1,000,000 |
| 2 | $\frac{\text{Number of facility department foremen}}{\text{Number of facility department craftsmen}}$ | Percent |

**TABLE 1.7 Equipment Control Sector**

| Number | Index | Units |
|---|---|---|
| 1 | $\frac{\text{Annual facility engineering cost by equipment type}}{\text{Replacement cost by equipment type}}$ | Percent |
| 2 | $\frac{\text{Preventive maintenance man-hours by equipment type}}{\text{Total facility department man-hours by equipment type}}$ | Percent |

**TABLE 1.8 Cost Control Sector**

| Number | Index | Units |
|---|---|---|
| 1 | $\frac{\text{Total facility department cost}}{\text{Facility value (replacement cost)}} \times 100$ | Percent |
| 2 | $\frac{\text{Cost of planning and scheduling}}{\text{Total facility department labor cost}} \times 100$ | Percent |

**TABLE 1.9 Overall Performance Sector**

| Number | Index | Units |
|---|---|---|
| 1 | $\frac{\text{Total costs due to facility department expenses}}{\text{Total sales of facility production}}$ | Cost per dollar |
| 2 | $\frac{\text{Value of lost production due to downtime}}{\text{Value of scheduled production}}$ | Percent |

## Departmental Management Audit Criteria

Management audit is a tool of management, instrumental in examining and determining the quality of performance. It is an instrument for measuring the effectiveness of a department's organization structure, its policies and practices, its systems and procedures, and its personnel. The proper use of this tool can be the means of equipping the department for better achievement of predetermined objectives. It can also do much, through review and appraisal, in finding improved methods of reducing costs and improving efficiency.

### Management Audit Checklist

A. Plans and objectives
   1. Have definite plans and objectives been established in the department?
   2. Are the plans and objectives in harmony with those of other departments as well as the organization as a whole?
   3. Has adequate time been allotted by those concerned with respect to forward planning and better ways of meeting objectives?
   4. Is there a clear understanding of objectives as to soundness and practicality?
   5. Is top management entirely in accord with the facility department's plans and objectives?
   6. What points should be considered in bringing about an improvement in the plans and objectives of the facility department?

B. Organizational structure
   1. Is there an organizational chart available and maintained current?
   2. Is the organization structure sound and effective?
   3. Does the organization reflect the program and objectives?
   4. Are the various duties and responsibilities delegated properly and defined clearly?
   5. Are the lines of authority effective from the standpoint of control?
   6. Is there any overlapping or duplication of functions?
   7. Can any organizational elements or functions be eliminated? Transferred to other departments?
   8. Can changes be made in the organizational setup to bring about increased coordination of activities?

9. Is there proper balance between the functions assigned to personnel?
10. Is there a lack of coordination or cooperation between the various functions?
11. Do the personnel concerned have sufficient understanding of responsibilities and authorities assigned?
12. What steps should be taken to increase the effectiveness of the organizational structure?
13. Does the average employee in the facility department have knowledge and understanding of the organizational structure?
14. Is there provision within the department for regular review of the organizational structure?

C. Policies, systems, and procedures
1. How are the facility department policies determined?
2. Have all facility department policies been reduced to writing?
3. Do the facility policies reflect the basic objectives and goals of management?
4. Are the facility policies positive, clear, and understandable?
5. Are the policies made known to the facility department personnel?
6. What provisions are set up to ensure compliance with established policies?
7. What is the policy pertaining to the selection of vendors and/or contractors?
8. Do the facility department purchasing requisitions reflect the necessary approvals by authorized personnel? Within dollar limits?
9. Are all facility department policies complied with?
10. Are the facility department systems and procedures meeting all current requirements and operating effectively?
11. Can the general routine in processing paperwork be improved?
12. Can improvements be made in the systems and procedures to bring about cost reductions?

13. Are the facility department policies, systems, and procedures reduced to writing?
14. Has sufficient consideration been given to internal control?
15. Have definite procedures been established to guide the conduct of all functions?
16. Are the procedures fully complied with?
17. Check for noncompliance with governmental laws, rules, and regulations.
18. Can any records and reports be eliminated?
19. What specific procedures require immediate study and revision?

D. Personnel
   1. What policies are established for selection, training, and assignment of personnel?
   2. Review the working conditions. What improvements are recommended?
   3. Is maximum use made of personnel? If not, what greater use can be made of personnel?
   4. Review the practices for the handling of personnel, including job descriptions, interviewing, placement, training, etc.
   5. What activities are in place for the development of personnel for promotion?
   6. Are new employees given sufficient orientation and training?
   7. What is the morale of facility department personnel and their attitude toward the organization?
   8. What is the rate of personnel turnover?
   9. Are there understudies for supervisory and key jobs?
   10. What is the status of absenteeism, sick leave, and requests for transfer?
   11. What percentage of routine clerical work is performed by supervisory personnel?

E. Layout and physical equipment
   1. Prepare a general layout of office and shops spaces and equipment.
   2. Is the office and shops laid out in a manner to obtain maximum utilization of space and efficient work areas?

3. What is the general condition of all systems and equipment?
4. Is equipment located for most optimum, extensive use?
5. Has provision been made for adequate storage spaces for parts, materials, and supplies?
6. Are the office and shop files reviewed regularly for transfer to storage? Records retention?

F. Operations and methods of control

1. What consideration has been given to the adequacy, clarity, and promptness of financial and management reports?
2. Is the normal lead time for processing non-emergency work orders generally adhered to?
3. Review the methods for scheduling regular work orders and preventive and predictive work orders?
4. What are the causes for overtime use and what can be done to eliminate them?
5. What are the principal means of work control?
6. How can the various operations be improved?
7. Can any operations and maintenance functions be eliminated, simplified, combined, or improved?
8. Are there any bottlenecks? What can be done to eliminate them?
9. What methods are established to measure productivity?
10. Are work units identified and performance standards developed? Are the performance standards attainable?
11. Is there need for work simplification training?
12. Are forecasts established to reflect future trends?
13. Is there budgetary control over all expenditures?
14. Do reports give comparison with past periods? With predetermined objectives?
15. Is there a means of ascertaining the cost variance on work order completions?
16. What is needed to increase the facility department's efficiency?
17. What can be done to increase the quality of work performed?

NOTES

Chapter

# 2

# Outsourcing Functions

## Recommended Tasks for Outsourcing

A facility's use of contract operations and maintenance functions can vary from a single operation, and/or maintenance, service contract, to a complete facility's operations and maintenance support contract, including supervision and engineering. In some cases, facilities use contract operations and maintenance as a way of obtaining needed contractors' skills and knowledge. Other functional operations and maintenance service tasks that are usually readily available, and useful to the facility manager, as *outsourcing* tasks items, are shown following: water treatment; custodial, pest control, trash disposal, and recycling; backflow preventers preventive maintenance; electrical distribution system preventive maintenance and testing; fired and unfired pressure vessel inspection and testing; landscaping, exterior and interior; elevator/escalator maintenance and repair; energy management control systems preventive maintenance, servicing, and programming; window washing; security; eddy current testing; and fire alarms, fire extinguishers, and sprinkler systems maintenance and testing. For a facility to have this full range of potential outsourcing operations and maintenance service tasks, it must be in a favorable geographical location, where such contractor capabilities exist to furnish these services. There are a growing number of such locations nation-wide today.

More rather than less involvement appears to be most attractive to new facilities, or ones that have undergone major renovations, since contracting all or a significant portion of their operations and maintenance services would not include changing existing systems. However, an established facility would not have a problem in this regard for it may be a useful facility engineering/management project to evaluate usefulness of existing systems to determine the need for change. Under these circumstances, the facility manager can develop his or her procedures during operations phase-in for new facilities; or during continued normal operations for existing facilities. From a practical standpoint, contract operations and maintenance services, in general, are more applicable to existing facilities, particularly if the facility manager is thinking in terms of total contract operations and maintenance services.

The facility manager's approach to outsourcing functional tasks may use one of the following alternative methods:

1. The facility manager can have his or her in-house operations and maintenance work force perform all operations and maintenance work, both day-to-day requirements and peak loads—with some craft specialty requirements contracted. Until recently, this approach was used by the majority of facility organizations in this country.

2. The facility manager can use his or her in-house operations and maintenance work force for performing all zday-to-day normal requirements, and bring in contractors' forces to handle major peaks—that is, emergencies, turnarounds, overhauls, and the like. In this case, the contractors' forces would be building trades craftsmen performing architectural/structural alterations, improvements, maintenance, or repair tasks; and/or HVAC contractors operations and maintenance craftsmen performing specific HVAC operations, maintenance, and repair tasks.

3. The facility manager can have his or her minimum in-house operations and maintenance work force perform only part of the day-to-day requirements, and can use contractors' operations and maintenance services to perform the balance of the day-to-day work, and to handle peak requirements.
4. The facility manager can elect to have no operations and maintenance working personnel assigned to his or her facility, and rely entirely on contractor forces to provide all operations and maintenance personnel for normal and peak operations and maintenance needs. This should be the exception, rather than the rule, for it has been found to be more efficient, effective, and economically beneficial to have a small number of in-house trades/craft personnel to provide HVAC operations and maintenance services, service calls performance, preventive and predictive maintenance procedures, and energy management control system operations and programming.

### Other considerations

To avoid possible labor union problems, the facility manager should give the facility's in-house work force all the work it can handle, including overtime work, before using contractors. The facility manager should review the labor union contract(s) the facility has with its own in-house operations and maintenance work force. By union contractual agreement, some facilities may not use contractors except for construction of major new facilities or major overhauls and turnarounds; others may have other specific restrictions that must be met regarding outsourcing of operations and maintenance service work.

The main objective, in this regard, is to accomplish the facility's required task, or tasks, on schedule, within cost limits, and without any disruption from the in-house or contractors' work forces. Every effort should be made to eliminate, or forestall, any disputes by notifying facility department personnel, and the various union committees, about the work planned to be performed by an outside contractor and why.

Such communication can minimize labor disputes and improve operations and maintenance personnel work attitude, and reduce friction when the contractor's personnel move into the facility to do their job.

**Motivating contractor workers.** When the facility manager elects to employ an outside operations and maintenance service contractor, the final contractor selection is usually based on reputation, experience, performance record, scope of services required, labor relations, and of course, price. Two often neglected areas that should also be considered in evaluating a contractor's ability to perform are: his methods of motivation and productivity measurement.

For contract operations and maintenance services to be profitable, the contractor must operate and maintain a client's facility more efficiently, and more economically, than can the client by using his own in-house work force. To accomplish this objective, the contractor must strive for the optimum utilization of his work force, which largely depends on proper motivation of his tradesmen/craftsmen, and measurement of their productivity, to ensure optimum output in accordance with the facility manager's contract performance work specifications.

## The Advantages and Disadvantages of Outsourcing

### Advantages

Outsourcing advantages are shown following:

1. The ability to fluctuate manpower use to exactly meet facility day-to-day needs permitting expenditure of less man-hours per year to perform a given amount of work. This method eliminates employment of full-time personnel to handle peak workloads and reduces the tendency for built-in featherbedding. It permits scheduling of seasonal work requirements to take advantage of seasonally good weather.

2. Seniority is not part of contract operations and maintenance services. Contractor tradesmen and craftsmen must compete with fellow workers to retain their jobs. In normal times, a job in the permanent nucleus of a contractor's operation and maintenance force is considered to be a choice job by most tradesmen and craftsmen. They usually get 40 hours per week of work, some overtime work, and, of course, area wage rates. They must produce to hold their jobs. In normal times, some facility managers feel that a facility can develop a more productive work force with contract labor than under average "facility" conditions using in-house labor.
3. A facility can meet the extraordinary manpower needs for start-up without over-manning for normal operations, and should work less overtime during start-up.
4. Since contractor tradesmen and craftsmen are already trade and craft trained, they require less training than the typical facility operations and maintenance employee to reach a given level of work performance efficiency.
5. Unlimited manpower permits scheduling of turnarounds, overhauls, alterations etc., to best fit overall facility economics and need. There is no requirement to attempt to fit these requirements to the availability of a given number of facility in-house employees. Equipment and systems downtime can also be shortened by using more personnel—to the extent practical and feasible.
6. If the facility manager elects, he or she can pass on to the contractor the responsibility for essentially all or a significant part of his or her facility's operations and maintenance services. This can leave more of the facility's key personnel with more time to concentrate on process operations—the making of more and better products, or providing better organization services. This does not mean that operations and maintenance services will be overlooked, because the contractor performing the services has a real economic incentive to

do a better job because, usually, operation or maintenance service contracts are an integral part of his annual income.

7. The facility manager can draw on the contractor for additional operations and maintenance services over and above normal facility operations and maintenance tasks. The contractor—if he or she is the right one—can provide additional supervision as needed for the provided services. The contractor can also provide intermittent supervision in specialty areas such as rigging, critical equipment inspection and repair, etc. The facility manager can use additional assistance from time to time for construction drawing takeoffs, work orders' planning and estimating, jobs sketching, equipment tests, etc., and still keep his or her staff at a constant level.
8. Another advantage often discussed is that the facility manager can reduce his or her investment in facilities, buildings, tools, and equipment. This is not a significant advantage. Over a period of time it has been observed that shops, storerooms, offices, etc. for contractors' personnel approach the norm used for the facility manager's in-house personnel. Shops of various types should be kept minimal. Larger, infrequently used equipment is available for lease, by the facility manager, from the contractor or others, regardless of "contract" versus "in-house" operations and maintenance requirements. There may be some savings on investment in tools if the contractor supplies them, but this does not present a significant savings.
9. The facility manager can shift essentially all responsibility for labor relations for operations and maintenance personnel to contractors, rather than be responsible for in-house personnel.
10. The facility manager can start with contract operations and maintenance; and if not satisfied with the results, he or she can phase into a facility in-house personnel performed operations and maintenance organization. If he or she starts with an in-house work force, there will

be great difficulties in making changes later on. If the facility manager is not satisfied with a contractor's performance, the contractor can be changed, which happens occasionally.

### Summary of advantages

The prime advantage of contract operations and maintenance services is that the facility manager has a "wide degree of flexibility," substantially more than that enjoyed with a typical facility in-house operations and maintenance work force.

### The disadvantages of contract maintenance

Facility managers should not examine only one side of any issue. The pitfalls (problems) must also be questioned and examined.

Some facility labor relations conditions permit the use of in-house mechanics who are not bound by trade or craft union jurisdictional lines in performing operations and maintenance work. Contractors, by and large, follow trade and craft union jurisdictional lines, except in emergency situations or where there is "right to work" laws in the state where the work is performed.

Trade and craft performance restrictions cannot be held as tight on operations and maintenance work as on construction work. On the other hand, unnecessary disregard for trade and craft performance restrictions should not be made. To avoid trouble, a contractor needs to exercise good judgment and consistency in trade and craft work assignments.

The degree of rigidity on trade and craft performance restrictions will vary from contractor to contractor, and from project to project. This is because different people are involved—different project managers, supervisors, craftsmen, tradesmen, etc. This is not the fault of the issued contract, but results largely from a lack of thorough training, good judgment, communication, and leadership on the part of certain contractor personnel.

With trade/craft performance restrictions, detailed work tasks planning is an absolute necessity. If the facility manager has all of the right people (the right trades and crafts), materials, tools, and equipment in the right place at the right time, he or she can have tradesmen and craftsmen working at their specialty without jurisdictional delays or squabbles.

On occasions, the "skilled" craftsmen and tradesmen are not so skilled, particularly when a geographical area is faced with an extremely high level of construction and facilities operations and maintenance opportunities. In some cases, facilities operations and maintenance personnel will be attracted back to construction work by substantial overtime opportunities, by offers of supervisory positions, and by special deals or incentives on the part of construction contractors. Facility managers contribute to this agenda by not controlling overtime and their contracts. A related problem has been that of staffing for substantial peaks. During periods of heavy construction work in a geographical area, normally the only way a contractor can attract substantial numbers of personnel, for facilities work, for relatively short periods of time, is to work overtime—enough to meet or exceed overtime that is being worked on construction projects, and/or pay higher than normal union or area hourly wage rates. Although the contractor and facility manager, for other considerations, often elect to work some overtime on turnarounds or overhauls, some feel that during high construction periods in their geographic areas, they work more overtime than normal to attract sufficient capable personnel to meet job demands. Thus, there is a conflict between construction, and contract operations and maintenance, due to unusual boom conditions in a geographic area. This situation will return to normal conditions, as the area construction level tapers off, and as training programs produce more skilled trade/craft personnel. Some facility managers feel that they cannot use contract operations and maintenance performance, for certain specialty areas, because contractors and cognizant unions cannot provide sufficient skills supervision, and tradesmen/craftsmen, with reasonable support backup. This has been true for instrumentation, particularly electronic, HVAC mechanics, and licensed stationary operating engineers. This, of course, does not preclude contract operations and maintenance use at a facility; but, it has been a deterrent. Accordingly, many facility managers have elected to develop their own work force with the necessary technical skills and cognizant licenses and certifications. Many contractors, and the unions concerned, are aware of this possibility, and similar problems, and are taking steps to provide these services.

Contract operations and maintenance wage rates are usually matched to construction rates in an area. On the average, construction wages have risen substantially faster than facility operations and maintenance wages, including fringe benefits. Most facility managers monitor this development carefully. They like the flexibility that goes with contract operations and maintenance; but, if the cost of contract labor gets too high in relation to facility inhouse labor rates, they cannot, or will not, be able to afford the need for contracting operations and maintenance functions.

### Contract services

A contractor can provide facility support services in many ways. He can offer general operation and maintenance services that include a review and audit of the facility department's organization; facility operations and maintenance work standards and procedures; custodial, pest control, trash removal, and recycling operations; response to emergency situations; procedures relating to the installation, operation, repair and maintenance of equipment and systems; facility inspection techniques, scheduling, record keeping, and reporting; operations, maintenance, and testing procedures for HVAC equipment and systems; and electrical distribution systems testing, maintenance, and repair. Facility inspections, as noted above, may extend to a complete facility structural examination, including rebar detection in concrete, underwater examinations, examination of timber and wood construction, load test certification, and corrosion studies. Inspection methods may include radiographic and ultrasonic, magnetic particle, and liquid penetrate testing; and visual examination.

Many contractors' shops are equipped with all the required tools and equipment to place the facility's equipment and systems back in service in the shortest possible time. There are, however, times when the equipment requiring repairs is so large that it would be difficult to move it out for repair. When the size of the equipment, its weight, or its stationary position prohibits such removal, on-site maintenance and/or repair may be the only option. In such cases, the contractor will dispatch his or her personnel to the facility to evaluate the problem, and discuss with the facility manager possible maintenance or repair procedures. When the facility manager has selected the most practical maintenance or repair option, the contractor will provide the necessary supervision, labor, materials, parts, equipment, and tools to complete the maintenance or repair at the facility.

A temporary contract arrangement, or one on a scheduled basis, can provide the additional personnel the facility manager requires to handle emergencies, or service breakdowns, of company mission impacting equipment; i.e., if the equipment or system fails, the company has to halt production. Such an arrangement permits the in-house operations and maintenance work force to meet the facility's daily needs, without concern for peak load conditions.

Each facility manager must carefully evaluate the abilities of his or her personnel to determine whether they can properly perform the equipment operations and maintenance tasks involved under normal conditions, and during possible emergencies, without any serious production losses. In the evaluation process, the facility manager should also compare and evaluate the costs of doing the work in-house to the cost of having it performed by a contractor.

### Quality Control

Conclusions on work quality by contractors are listed following:

1. Contract operations and maintenance work force performance offers the facility manager an opportunity to obtain a work force that produces good quality work. When the facility manager takes the opportunity to evaluate and eliminate contractors, who normally perform less than optimum quality work, the provided work quality will usually be of excellent quality. Where no screening is performed, work quality provided will generally be less than satisfactory.
2. Work quality provided depends on the quality of supervision provided. Good supervision is critical to obtaining quality work, yet this is an important factor rarely considered. However, the quality of project supervision, if less adequate than the contractors had promised, will normally result in low-quality work.
3. Contract operations and maintenance work performance is no guarantee of delivered quality work. Tradesmen and craftsmen hired by contractors are not better than those recruited for the facility's in-house work force. They lack the specific experience that the regular employees acquire in time at the facility, but they often compensate for this by bringing a greater breadth of experience to the job.

### Flexibility

The biggest advantage of contract maintenance is flexibility. The ability to recruit, and dissolve, a large work force quickly is most often cited as the contract operations and maintenance major advantage. Other advantages are

1. Contract operations and maintenance work are applicable to more than rebuilds, turnaround, or alterations work.

2. It is difficult to institute a reduction-in-force for an in-house work force.

### Reduced capital expenditures

In these competitive and rapidly changing times, private and public sector organizations are all deeply concerned with either maximizing profits or making a major improvement in services delivery, i.e., making as high a return on investment as possible. When an organization for profit (private sector) is reduced to its simplest operating concept—profit—profit can be increased. Public sector organizations' simplest operating concept is to provide required quality services—optimum delivered services—at the least overall cost (see the following list).

1. *Private sector.* Higher production and in turn higher sales, and higher average prices for products or services—in other words, more income.
2. *Public sector.* Meet required services delivery specifications with the required services quality.
3. *Public and private sectors.* Hold the line on, or reduce operations costs of, any and all kinds less operations costs.

Facility operations and maintenance performance can have significant effects on the above items. Facilities kept in good condition can provide higher product output per unit of time and can also enjoy a higher service or use factor, both adding to increased production and in turn high sales—thereby producing more income. Efficient and effective operations and maintenance work performance can reduce direct operations and maintenance costs, and therefore reduce the organization's overall operations costs.

Private and public sector facility managers, in seeking new ideas for improving the quality of operations and maintenance tasks performance and reducing operations and maintenance costs, have designed and developed the concept of a comprehensive outsourcing (contract) operations and maintenance program to achieve these objectives. From a cost/benefit comparison viewpoint, it would be difficult to prove that contract operations and maintenance usage are less expensive than in-house performance due to the lack of availability of good cost data in most facility organizations.

Engineering judgment, and a review of many facility contract operations and maintenance work experiences, reveal that:

1. There is little "make work" in contract operations and maintenance services performance.
2. Contractors' profits were not too high.
3. Less money is spent on "frills" than does an in-house work force.
4. Contractors' wage rates are not too high indicating a supposition that other available savings will offset the wage differential between in-house and contractor employees.

#### Employing specialty contractors

What are the procedures that should be used for using contract operations and maintenance services? In other words, what actions and thought processes should a facility manager use from the time he or she decides he or she wants to consider contract operations and maintenance services support until a contract is awarded and the contractor is onboard performing the services. In logical sequence these actions are:

1. Most facility managers usually have background and experiences for managing in-house operations and maintenance work performance, but need to gather comparable information and data on past contract operations and maintenance work performance. The best sources for this information are other facility managers currently using operations and maintenance contractors.

2. After as much information and data as possible is gathered, a careful analysis and evaluation will determine what outsourcing task is really applicable in the individual facility's case.
3. Next, facility managers must make a comparison of the advantages offered by contract operations and maintenance with using his or her own in-house work forces. However, caution, and complete objectivity, must be exercised because intangibles have to be considered and assumptions made. The analysis made must consider the overall effect on the facility, as follows, during the contract: start-up (phase-in) period, operational period, and the close-down (phase-out) period after the contract has expired, or has not been renewed.
4. If after an overall evaluation of costs and benefits has been made, it is concluded that contract operations and maintenance services offer distinct advantages, the facility manager should then further refine the evaluation, and decide on the procedure to be followed. He or she should also decide who will perform work planning and scheduling, the extent the contractor will be involved in the facility's cost accounting services, whether the contractor will be have to provide and staff support services such as purchasing, drafting, reporting, record keeping, etc.

The defined tentative contractor operational procedures should be established so that when taking bids, or negotiating with a contractor, the general scope-of-work and performance work statement can be described. It will also be useful to consider the type and size of contractor, and in-house, organizations required to manage the contract.

Contractor selection should be made based on a projected long-term contractual arrangement. The contractor and the facility manager must work as a team in executing the awarded contract. The following major points should also be considered in this regard:

1. The contractor's integrity.

2. The contractor's experience. Has he performed operations and maintenance work with his staff, or has he been a labor broker with facility managers shouldering the responsibility?
3. What is the extent and depth of the contractor's experience?
4. What does the contractor offer in terms of supervision, and trade and craft personnel skills? This is a most important consideration.

The following criteria must be carefully reviewed as regards the contractor's capability.

- Does he or she have an adequate staff to support project work forces? Are these personnel experienced on contract operations and maintenance projects?
- Who does he or she offer as his or her project manager? Has this person been in the contractor's employ prior to this contract for at least three (3) years? Does the contractor really know the proposed project manager's capabilities? Is he or she experienced and skilled in supervising operations and maintenance work?
- Can the contractor offer sufficient skilled assistant supervisors to manage the job? What are their experience and background? Will he or she have to hire unknowns to staff the project? Can the contractor offer adequate supplemental and/or specialty supervisors?
- What is the contractor's situation with regard to any unions involved—particularly at the local level?
- Does the contractor have complete knowledge of the individual tradesperson's and craftsperson's basic skills, special skills, and capabilities?
- Are his or her company and key people respected by the various unions and their members?
- Is the contractor inclined to manage in short-term expediencies, or does he or she adhere to the formal labor agreements? Does he or she obtain what he or she is entitled to

under his or her labor agreements, but still deal fairly with union members?

- How complete is the contractor's service in his specialty? Do his or her services match your needs? If not, are you willing to split work responsibility? Does he or she have necessary equipment, tools, and facilities to support your operation?

It is difficult for the facility manager to obtain complete and detailed answers to these questions. Some contractors are prone to exaggerate when discussing their capabilities. The facility manager needs to probe deeply and take a "show me" attitude. Be sure to obtain the true story before drawing any conclusions.

One good measure of a contractor's capability is his overall effectiveness. Has he or she been able to consistently perform repeat work for facility managers in the area? Does he or she receive a fair share of the area's contract work? Does he or she make reasonable profits on his contracted work? Facility managers should request financial information from contractors, not only to be sure the contractor is fiscally sound, but also to measure the company's progress in terms of profits. The ability to consistently make reasonable profits is conclusive evidence of a well-managed organization.

### Selecting an outsourcing source

Several approaches are open to facility managers in contractor selection:

1. One approach is to check with other facility managers in the area to determine what contractors they use and the degree of success they are having. Or, he or she may obtain information of experiences of other facility managers at monthly meetings, seminars, etc.

2. Another approach is to contact the manufacturer of some of the more sophisticated equipment currently operating in the facility. The manufacturer may provide maintenance and/or repair services for his equipment, or may be able to suggest a contractor for that service. Many manufacturers of

electrical equipment (motors, generators, transformers, and switchgear), boilers, air-conditioning units, and temperature controls (calibration) have service shops throughout the country and can provide such a service on a contract basis, or on an "as needed" arrangement. Any production or facility equipment that would cause the facility to shut down would be of primary concern to the facility manager. If such equipment is beyond the skills and abilities of in-house personnel, the facility manager should determine the types of services that are provided by the equipment manufacturer.

Many engineering and construction firms provide an operations and maintenance contract service for a facility's operating equipment. Although such services are offered mainly for refineries, chemical plants, and power generating stations, they are also available for any type of public or private sector facility. These firms can supply specialists in planning and scheduling, rotating equipment, piping, rigging, and electrical equipment. They are capable of handling all equipment operations, maintenance, emergency repairs, capital improvements, overhauls, turnarounds, and other required services. Some of their systems incorporate sophisticated record keeping and maintenance management (CMMS) procedures to include preventive and predictive maintenance programs, and emergency work procedural techniques, to ensure that all work is performed logically and efficiently.

One other approach would be to examine a directory containing the names of various contractors' associations to determine those with specific specialties interest. The associations will provide the names of contractors you are interested in. Also look for contractors' names in the local telephone book. Follow-up telephone calls can determine if further correspondence, or contact, with those companies is warranted.

### Judging abilities of an outsourcing firm

How can the facility manager judge a contractor's work quality and technical ability? There are eight important characteristics to look for prior to obtaining services of an operations and maintenance contractor: technical expertise of personnel, local support availability, service reputation, pricing policy, business experience, financial stability, equipment repair facilities (including availability of tools and materials), and speed of response to service calls.

Besides the facility manager, the operations supervisor, maintenance supervisor, engineering supervisor, and/or electrical and mechanical engineers should be involved in the contractor selection process.

Annual preventive maintenance agreements, a 2-or-more-years' preventive maintenance agreement, and service on an "as needed" basis are the types of contracts normally selected. Other types would be for special frequency on specific equipment and for emergencies.

It is important that the facility manager include in the service contract all details related to the equipment and systems being serviced. The facility manager and the contractor must agree to all performance work statement specifications and the contract costs.

One major consideration to be evaluated before signing a service contract is to ensure that the contractor has adequate insurance coverage, and that the insurance information and the amount of coverage involved become part of the contract. Accidents that occur in the facility as a result of the contractor's personnel actions should be covered by his insurance. Contractors must carry ample insurance in the following areas so that the facility is not held responsible for any accidents to the contractor's or company's employees or damage to facility equipment: workmen's compensation and occupational disease and protective liability insurance.

It may be advisable for the facility manager to have a protective liability insurance policy to protect the facility from subcontractors that may be used, but are not covered under the prime contractor's insurance policies.

The contract awarded should guarantee for the life of the contract that the contractor remedy failures, at no additional cost to the facility manager, due to poor workmanship and/or defective materials furnished by the contractor.

A penalty clause in the contract, to protect the company, is advisable if an equipment or system installation has to be completed by a deadline. The penalty clause can serve as an incentive and also protect the company against costly delays. However, it may be detrimental in pushing the contractor to rush the installation, and result in poor workmanship being provided. The final decision as to whether or not to include, or exclude, a penalty clause should be based on the contractor's reputation and the criticality of the project.

No matter what type of service contract is selected, the facility manager should assign one of his or her engineering or supervisory staff to review, periodically, all contractor submitted reports. With such an arrangement, it is possible to monitor the progress of the work being performed, and to call attention quickly to tasks work performance, or costs, that may be deviating from the contracted norms.

### Responsibility for outsource firm administration and control

Since, as shown above, the "contract document" is considered a key to contractor administration and control, it is necessary to ensure that this document provides details on every aspect of the work to be performed and/or materials and parts to be provided and the standards to which the work will be performed. In addition to the controls provided by a well-defined contract, the supervisor of each section that will use contractor services should be delegated operational control and tasked to complete a monthly evaluation report such as that shown in Fig. 2.1, a monthly evaluation report for contractors, for each facility contractor.

### Outsource contract administration

Viable and effective outsource agreements require an extremely high degree of specificity in every phase. Certainly, that concept may be applicable to any form of contract. However, the very nature of the outsource agreement, i.e., reliance on "outside" maintenance and operations personnel, not employed on a full-time basis, and thereby beyond routine facility management controls, makes it

**Monthly Evaluation Report for Contractors**

| | Performance Factor | Raw Score 0-100 | Weight Factor | Weighted Score (points) |
|---|---|---|---|---|
| 1. | Contractor has demonstrated flexibility and responsiveness to overall support. | ______ | X 0.10 | ______ |
| 2. | Contractor has provided required planning and scheduling of resources to meet needs. | ______ | X 0.10 | ______ |
| 3. | Contractor's organization is adequately staffed and properly supervised. | ______ | X 0.10 | ______ |
| 4. | Contractor has met all reporting requirements in timely and accurate manner. | ______ | X 0.10 | ______ |
| 5. | Contractor has identified potential problems and provided alternative solutions. | ______ | X 0.10 | ______ |
| 6. | Contractor's personnel are qualified for assigned positions. | ______ | X 0.10 | ______ |
| 7. | Contractor has provided efficient utilization of equipment and facilities. | ______ | X 0.10 | ______ |
| 8. | Contractor has responded to previously noted problems or deficiencies. | ______ | X 0.10 | ______ |
| | | | **Total Weighted Score** | ______ |

**Remarks:** ______________________________

______________________________

______________________________

**A Raw Score of 70 or above = Acceptable Performance**
**A total Weighted Score of 49 or below = Unacceptable Performance**
(Maximum Weighted Score Possible is 80 points)

**Figure 2.1** Monthly evaluation report for contractors.

critical that every phase of the contract be specific enough, so as to avoid any misunderstandings. Nonetheless, the onus of such restrictions should not deter the Facility Manager from seeking outside contracting. Again, various mechanical and technical disciplines simply cannot be delegated "in-house," without sacrificing safety and efficiency.

### Negotiating the contract

The initial step in negotiating the contract occurs well before the final contractual arrangement is put to paper. More specifically, the Facility Manager must attempt to articulate his or her needs, before initiating negotiations with any contractors. In fact, it would be good practice to detail such requirements in a written prospectus, whether in the form of a bid proposal or by way of a request for contractual services. However, in either case, the key element is the provision of complete notice to the prospective contractor, of the Facility Manager's expectations. Further, any such document must ultimately be incorporated into the final contract, either as an Addendum or in some other form.

In determining pricing requirements, the facility manager should always look to his or her costs for comparable in-house services, as a baseline. Obviously, specialists, such as elevator maintenance personnel, must be able to perform such services less expensively, or at least be able to accomplish such services more efficiently then in-house personnel.

The prospectus, or bid proposal, must require references and suggested maintenance and operation plans by each bidder. These written plans are necessary so that accurate comparisons can be effected between bidders. Again, such written submissions must be included in the written contract.

### Specification of services; responsibility for communications; supplies, materials, equipment, and utilities

Once the facility manager has determined the particular services that must be supplied by a contractor, those services must be written into a *Support Services Agreement.* Accordingly, it becomes paramount to allocate responsibility between the parties. In other words, there should be a confirmable understanding as to each party's duties and responsibilities: e.g., who is expected to supply materials, equipment, and utilities; which costs are included in the Agreement; and the extent of labor included.

### Certified outsource personnel to be used

It is not enough for the facility manager to rely on a contractor's reputation with regard to a particular field of endeavor. Better practice is to include language in the Agreement that requires that the nature and/or level of certification for each employee that will be utilized on-site reach prescribed levels in accordance with the laws, codes, and regulations of the jurisdiction in which the work is to be performed. Moreover, it is not unreasonable to require that only those individuals qualified, or certified, to perform the *particular services specified in the outsource agreement* be utilized.

### Scheduling work and/or reporting requirements

Every outsource agreement must include a prescribed basic maintenance and operation schedule, along with a system of reporting the level that maintenance and operations have been accomplished, in accordance with the agreement. There are a variety of methods available for such reporting requirements. However, it is suggested that the operative phrase should be "the simpler the better."

Accordingly, the outsource agreement should include a "check-off" list and/or schedule of required contacts with appropriate supervisory personnel, to suffice as reporting requirements.

It is also important to specify in any contract what is *not* to be accomplished. In other words, the basic scope of any outsource agreement is invariably, as stated above, a listing of what services the Facility Manager expects to be performed, for the fees to be paid. However, to avoid any misconceptions in terminology or expectations, it is good practice to specify what services will *not* be performed; what parts and/or supplies will *not* be provided by the contractor (e.g., whether worn-out parts are included in the contract); and/or what assistance or notification will *not* be required of Facility Manager.

### Supervision of contractual execution

In any outsource agreement the supervision of contract execution must be relegated by in-house personnel. While this statement is simple in nature and, perhaps, logical in essence, nevertheless, it is good practice to specify the nature and scope of supervision, to avoid any potential problems.

By delegating Facility Manager's responsibility for supervision of *each* particular contract duty, the facility manager must ensure that a manager, or management team, will develop sufficient expertise to understand and evaluate the technical services. By the same token, this provides the contractor with a contact person, or committee, for emergency actions and with respect to any modification on the provided services.

### Conclusion

The traditional adage "good fences made good neighbors" certainly applies in the outsource contract agreements. Certainly, any contract may have to be modified over its life. However, by crafting a specific and detailed agreement you avoid communication problems, failure of service, and, hopefully, litigation. (See Figs. 2.2 and 2.3 for samples of well-written outsource agreements. See also Barbara Czegel, *Running an Effective Help Desk: Planning, Implementing, Marketing, Automating, Improving, Outsourcing,* Wiley/QED, New York, 1994; and Donald K. Hicks, John H. Williamson, and Carl G. Lewis, *Guide to Preparing Work Descriptions for Performance Work Statements for Contracted Maintenance Activities for Army Installation Directorates of Logistics,* U.S. Army Corps of Engineers, Construction Engineering Research Laboratory, Champaign, Ill., 1991.)

### Other outsourcing considerations

One result of the recession of the 1990s is recognition of the value of managing facilities as an asset. As soon as a facility is constructed it begins to deteriorate. Clearly, funds that must be used for operations and maintenance

**SUPPORT SERVICES AGREEMENT**

TO:

PROPOSAL No. 01MP95009
DATE SUBMITTED: August 14, 1994
PHONE No.:

AGREEMENT No.:

Location of Agreement: 11510 Falls Road

Siebe Environmental Controls agrees to provide the services described in the attached schedules in accordance with the following terms and conditions.

| Schedules Attached | Type of Service |
|---|---|
| A | Temperature Control |
| B | Mechanical System |
| E | Filter Service |
| G | Additional Services |
| H | List of Equipment |

**Terms and Payment:**

We agree to furnish the services as described on the schedules checked above for the annual price of $7,440.00
SEVEN THOUSAND FOUR HUNDRED FORTY DOLLARS**************************************************************

This Siebe Environmental Controls Service Agreement shall begin on the 15 day of August, 1994 and shall continue for a period of one year(s) and from year to year thereafter until terminated. Either party may terminate this agreement upon thirty days written notice prior to the anniversary date of the agreement.

Invoices will be issued Monthly (XX) Quarterly ( ) Semi-Annually ( ) Annually ( ) Special ( ) as agreed. Payment will be made in 30 days.

The contract price shall be subject to adjustment yearly to recognize any changes in costs. Notice of proposed adjustments to the annual price will be provided to you at least sixty days prior to agreement renewal date.

SIEBE ENVIRONMENTAL CONTROLS

Company: ____________ Submitted By: Doug Wiggins

Accepted By: ____________ Accepted By: ____________

Title: ____________ Title: ____________

Date: ____________ Date ____________

**Figure 2.2** Support services agreement.

could otherwise be directed to increase an organization's growth and/or profit. It is the responsibility of an organization's facility maintenance organization to preserve an organization's capital assets.

Facility management is a diverse complex function. To satisfactorily accomplish all the required tasks internally would take a large, diverse, and highly trained staff. Very small facilities cannot afford to provide this required level of staffing. Even most large facilities usually cannot justify staffing in-house specialists that are required only occasionally. Outsourcing allows an organization to expand its capabilities and resources without the need to expand its work

**General Conditions:**

Preventive maintenance will be performed during normal working hours and are defined as 7:30 AM 4:00 PM Monday through Friday inclusive, excluding holidays.

Reasonable means of access to the equipment being maintained shall be provided by the owner. Our service does not include the normal operation of your system such as starting, stopping, or resetting of the equipment described. However, Siebe Environmental Controls shall be permitted to start and stop all equipment necessary to perform the herein agreed services.

Siebe Environmental Controls shall not be liable for any loss, delay, injury or damage that may be caused by circumstances beyond its control including, but not restricted to acts of God, war, civil commotion, acts of government, fire theft, corrosion, electrolytic action, floods, lightning, freeze-ups, strikes, lock-outs, differences with other trades, riots, explosions, quarantine restrictions, delays in transportation, shortage of vehicles, fuel, labor materials or malicious mischief.

Siebe Environmental Controls' responsibility for injury or damage to persons or property that may be caused by or arise through the maintenance service, or use of the system(s) shall be limited to injury or damage caused directly by our negligence in performing or failing to perform our obligations under this agreement. Siebe Environmental Controls shall not be responsible for the failure of furnishing labor or material caused by reason of strikes or labor troubles affecting our employees who perform the service called for herein, or by unusual delays beyond our reasonable control in procuring supplies or for any other cause beyond our reasonable control. In no event shall Siebe Environmental Controls be liable for business interruption losses or consequential or speculative damages.

Siebe Environmental Controls will present a service report for your signature at the completion of each visit.

We will not be required to make safety tests, install new attachments or appurtenances, add additional controls, and/or revamp or renovate existing system with devices of a different design or function to satisfy conditions established by insurance companies, laboratories, governmental agencies, etc.

In the event the system is altered, modified, changed or moved, Siebe Environmental Controls reserves the right to terminate or renegotiate the agreement based on the condition of the system after the changes have been made.

If emergency service is included in this agreement and is requested at a time other than when we would have made a scheduled preventive maintenance call, and inspection does not reveal any defect required to be serviced under this agreement, we reserve the right to charge you at our prevailing service rates.

Siebe Environmental Controls reserves the right to discontinue this maintenance service agreement at any time, without notice, unless all payments under this contract shall have been made as agreed.

If replacement of parts are included in this agreement, it is understood that Siebe Environmental Controls will not be responsible for the replacement or repair of boiler tubes, boiler sections, boiler refractory, chimney, breeching, refrigeration evaporators, refrigeration condensers, water coils, steam coils, concealed air lines, fan housings, ductwork, water balancing, decorative casings, equipment painting, disconnect switches, electrical power wiring, water, steam, and condensate piping or other structural or non-moving parts of the heating, ventilation, and air conditioning systems. Replacement control valves and dampers, when in our judgment they are required, are covered by this agreement. However, removal from and reinstallation to pipes and ductwork is not included. Siebe Environmental Controls will not be required to make replacements or repairs necessitated by reasons of negligence, misuse or other causes beyond our control except ordinary wear and tear.

If equipment becomes non-repairable due to unavailability of replacement parts, Siebe Environmental Controls will no longer be required to maintain or service such equipment as a part of this agreement. However, Siebe Environmental Controls will assist the owner in replacing the equipment at prevailing service rates.

It is agreed that the equipment, piping, ductwork, controls, etc., have been installed basically as shown on the contract drawings for this building and that the installation and performance of these systems is acceptable to the owner.

It is further understood that the equipment covered under this agreement is in maintainable condition and eligible for a maintenance agreement. If at the time of start-up or on the first inspection, repairs are found necessary, such repair charges will be submitted for the owner's approval. If these charges are declined, those items will be eliminated from the agreement and the price of the agreement will be adjusted in accordance with equipment covered.

**Figure 2.2** *(Continued)*

force. "Outstanding" means purchasing services by contract from vendors. "Outsourcing" involves contracting for a contractor's time and effort rather than for a specific end product. These contracts are typically referred to as *service* contracts.

Virtually all or part of a facility management organizational functions can be outsourced. In addition, several of these functions can be provided by a single vendor.

SCHEDULE A

AUTOMATIC TEMPERATURE CONTROLS

Scheduled maintenance inspections shall be performed during normal working hours

Frequency:

() Weekly
(x) Quarterly
() Annually
() Other ____________
() Monthly
() Semi-Annually

Coverage:

() Scheduled Maintenance Only
(x) Scheduled+Unscheduled Maintenance
() Repair Materials
(x) 24-Hour Response

Equipment Covered: All electric and pneumatic controls

**Air Compressor**
Drain Tank and Inspect
Note Condition of Supply Air
Check Run Time
Check Pulley and Belt Tension
Check PRV and Filter Station
Change Oil (Annually)

**Air Handling Units**
Review Sequence of Operations
Inspect and Lubricate Dampers and Valves (Annually)
Check Pilot Positioners
Check Panel Mounted Controls
Check Transmitters and Gauges
Calibrate as Needed
Clean Panel Face
Check Auxiliary Devices (PE's EP's)
Check Safety Devices (Freeze, Fire)

**Air Drier**
Check Traps
Clean Condenser
Check Expansion Valve and Refrigerant

**Time Clocks**
Check Time Settings and Operations

**Terminal Equipment**
Check Room Thermostats (Annually)
Calibrate as Needed
Check Valves; Damper Actuators

**Water Systems**
Check Master/Submaster Controls
Calibrate as Needed
Check Valves, PE's, Step Controllers
Check Safety Controls

**Figure 2.2** *(Continued)*

The term outsourcing has become a catch-all for a wide variety of arrangements, for procuring operations and maintenance functions contractually. These range from the contracting out of individual services to hiring a vendor to take on a wide variety of facility management operations, maintenance, and support functions. It has become an accepted tool along with total quality management, and business reengineering, for dealing with changes in the facility management environment. Outsourcing is being used to achieve better customer service, competitive advantage, and lower costs.

According to a 1995 survey by the Association of Facility Engineers (AFE), nearly 70 percent of facilities professionals believe that downsizing and outsourcing will continue to increase. The Outsourcing Institute expects the market for outsourced facilities services surge dramatically in the next five years.

SCHEDULE B-5

**CONDENSING UNITS**

Scheduled maintenance inspections shall be performed during normal working hours.

Frequency:

() Weekly (X) Monthly
() Quarterly () Semi-Annually
() Annually
() Other ______________

Coverage:

() Scheduled Maintenance Only
(X) Scheduled+Unscheduled Maintenance
() Repair Materials
(X) 24-Hour Response

Equipment Covered: See Equipment Schedule H

**Air Cooled - Startup Inspection**

1. Review manufacturer's recommendation for startup.
2. Energize crank case heater per manufacturer's recommendation for warmup.
3. Remove all debris from within and around unit.
4. Visually inspect for leaks.
5. Check belts, pulleys and mounts. Replace and adjust as required.
6. Lubricate fan and motor bearings per manufacturer's recommendation.
7. Inspect electrical connections, contactors, relays and operating/safety controls
8. Check motor operating conditions
9. Check and clean fan blades as required.
10. Check and clean coil. Straighten fins as required.
11. Check vibration eliminators. Replace or adjust as required.
12. Check compressor oil level, acid test oil and meg hermetic motor. Change oil and refrigerant filter drier as required.
13. Check and test all operating and safety controls
14. Check operating conditions. Adjust as required.

**Mid-Season Inspection**

1. Visually inspect for leaks.
2. Lubricate fan bearings per manufacturer's recommendation.
3. Lubricate motor bearings per manufacturer's recommendation.
4. Check belts and sheaves. Replace and adjust as required.
5. Clean and straighten fins as required.
6. Check operating conditions. Adjust as required.

**Air Cooled Seasonal Shutdown**

1. Review manufacturer's recommendation for shutdown.
2. Verify economizer operation (if applicable).
3. Visual inspection of leaks.
4. Verify auxiliary heater operation (if applicable).
5. Inspect belts and filters.
6. Test all safety controls

**Optional to Above Procedures**

() 1. Clean coil with chemical and high pressure spray. (See Water Treatment)

**Figure 2.2** *(Continued)*

The key benefits of outsourcing facility operations and maintenance functions, according to Robert Dickhaus, COO of Johnson Controls World Services, Cape Canaveral, Florida, include:

- Time to focus on core business
- Accelerated re-engineering
- Technology infusion
- Shared risks
- Extended technical and management resources

Water Cooled - Startup Inspection

1. Review manufacturer's recommendation for startup.
2. Energize crank case heater per manufacturer's recommendation for warmup.
3. Visually inspect for leaks.
4. Vent system of trapped air.
5. Inspect electrical connections, contactors, relays and operating/safety controls.
6. Check vibration eliminators. Replace or adjust as required.
7. Check compressor oil level, acid test oil and meg hermetic motor. Change oil and refrigerant filter drier as required.
8. Check and test all operating and safety controls
9. Check operating conditions. Adjust as required.

Mid-Season Inspection

1. Visually inspect for leaks.
2. Check operating conditions. Adjust as required.

Air Cooled Seasonal Shutdown

1. Review manufacturer's recommendation for shutdown.
2. Verify economizer operation (if applicable).
3. Visual inspection for leaks.
4. Verify auxiliary heater operation (if applicable).
5. Inspect belts and filters.
6. Test all safety controls.

Optional to Above Procedure

() 1. Inspect and punch condenser tubes.

Preseason Inspection

1. Inspect fireside of boiler and record condition.
2. Brush and vacuum soot and dirt from flues and combustion chamber.
3. Inspect firebrick and refractory for defects. Patch and coat as required.
4. Visually inspect boiler pressure vessel for possible leaks and record condition.
5. Disassembled, inspect and clean low-water cutoff.
6. Check hand valves and automatic feed equipment. Repack and adjust as required.
7. Inspect, clean and lubricate the burner and combustion control equipment.
8. Reassemble boiler.
9. Check burner sequence of operation and combustion air equipment.
10. Check fuel piping for leaks and proper support.

Seasonal Startup

1. Review manufacturer's recommendations for boiler and burner startup.
2. Check fuel supply.
3. Check auxiliary equipment operation.
4. Inspect burner, boiler and controls prior to startup.
5. Start burner, check operating controls. Test safety controls and pressure relief valve.
6. Perform combustion test and adjust burner for maximum efficiency.
7. Log all operating conditions.
8. Review operating procedures and owner's log with boiler operator.

Season Shutdown

1. Review owner's log. Log all operating conditions.
2. Shut off burner and open electrical disconnect.
3. Close fuel supply valves.
4. Review boiler operation with boiler operator.

Monthly Operating Inspection

1. Review owner's log. log all operating conditions.
2. Inspect boiler and burner and make adjustments as required.
3. Test low water cutoff and pressure relief valve.
4. Check operating and safety controls.
5. Review boiler operation with boiler operator.

**Figure 2.2** *(Continued)*

SCHEDULE B-8

FANS & CENTRAL FAN SYSTEMS

Scheduled maintenance inspections shall be performed during normal working hours.

Frequency:

() Weekly
() Quarterly
() Annually
() Other ____________
(X) Monthly
() Semi-Annually

Coverage:

() Scheduled Maintenance Only
(X) Scheduled+Unscheduled Maintenance
() Repair Materials
(X) 24-Hour Response

Equipment Covered: See Equipment Schedule H

CENTRAL FAN SYSTEMS - ANNUAL INSPECTION

1. Check and clean fan assembly.
2. Lubricate fan bearings per manufacturer's recommendations.
3. Lubricate motor bearings per manufacturer's recommendations.
4. Check belts and sheaves. (Replace and adjust as required.)
5. Tighten all nuts and bolts.
6. Check motor mounts and vibration pads. (Replace and adjust as required.)
7. Check motor operating conditions.
8. Inspect electrical connections and contactors.
9. Lubricate and adjust associated dampers and linkage.
10. Check fan operation.
11. Clean outside air intake screen.
12. Check and clean drains and drain pans.
13. Check and clean strainers, check steam traps and hand valves.
14. Check filter advancing mechanism. Lubricate and adjust as required.
15. Inspect filters.
16. Check heating and cooling coils.
17. Inspect humidifier.

SEMI-ANNUAL INSPECTION

1. Lubricate fan bearings per manufacturer's recommendations.
2. Lubricate motor bearings per manufacturer's recommendations.
3. Check belts and sheaves. (Replace and adjust as required.)
4. Clean outside air intake screen.
5. Check filter advancing mechanism. Lubricate and adjust as required.
6. Inspect filters.
7. Check heating and cooling coils.
8. Check humidifier.

**Figure 2.2** *(Continued)*

For outsourcing to be successful, management must have clear goals and objectives. When a decision is made to outsource existing functions in the organization, management must be sensitive to and address the anxiety this may create among its workers. Management should make a good faith effort to assist its employees who may be displaced by outsourcing find other employment. Frequently, the outsource provider will provide employment opportunities for these individuals. However, if the main goal of outsourcing is a reduction in staff, then management must prepare itself for the gloom which is likely to permeate the work force.

Outsourcing offers many potential benefits if the associate contract is structured and managed properly.

SCHEDULE B-9

## PUMPS

Scheduled maintenance inspections shall be performed during normal working hours.

Frequency:

| | | | | | | |
|---|---|---|---|---|---|---|
| 0 | Weekly | (X) | Monthly | | 0 | Scheduled Maintenance Only |
| 0 | Quarterly | 0 | Semi-Annually | | (X) | Scheduled+Unscheduled Maintenance |
| 0 | Annually | | | | 0 | Repair Materials |
| 0 | Other | | ________ | | (X) | 24-Hour Response |

Coverage (right columns above).

Equipment Covered: See Equipment Schedule H

ANNUAL INSPECTION

1. Lubricate pump bearings per manufacturer's recommendations.
2. Lubricate motor bearings per manufacturer's recommendations.
3. Tighten all nuts and bolts. Check motor mounts and vibration pads. (Replace and adjust as required.)
4. Visually check pump alignment and coupling.
5. Check motor operating conditions.
6. Inspect electrical connections and contactors.
7. Check and clean strainers and check hand valves.
8. Inspect mechanical seals. Replace as required OR inspect pump packing. Replace and adjust as required.
9. Verify gauges for accuracy.

SEMI-ANNUAL INSPECTION

1. Lubricate pump bearings per manufacturer's recommendations.
2. Lubricate motor bearings per manufacturer's recommendations.
3. Check suction and discharge pressures.
4. Check packing or mechanical seal.

Figure 2.2 *(Continued)*

Outsourced services may cost less because the contractor may be able to pay its employees 15 percent more or less in wages than what the current in-house staff is being paid. Vendors usually have a large pool of skilled labor and supervisors so that they are able to satisfy overtime, turnover, and vacancy requirements. The organization can also avoid the costs that are associated with hiring, training, scheduling, and disciplining in-house personnel. Similarly the vendor can reduce the facility's exposure to Worker's Compensation. In addition, vendors because of their experience and size are often able to obtain better insurancerates and to absorb services on the job accidents and injuries. Large vendors often can provide their skilled

SCHEDULE B-10

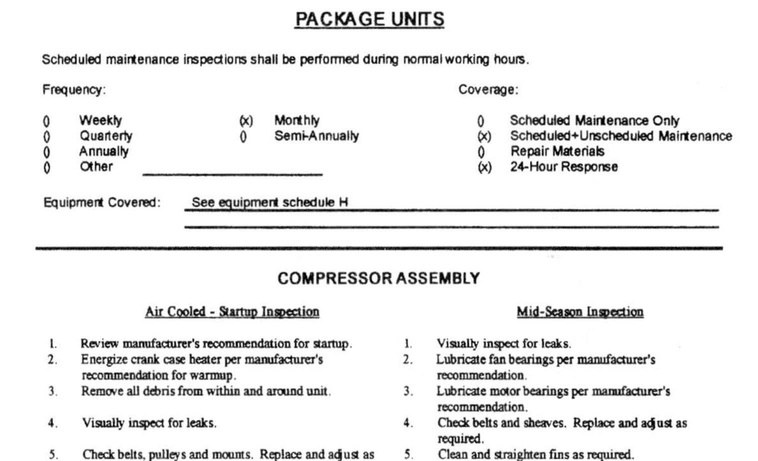

**PACKAGE UNITS**

Scheduled maintenance inspections shall be performed during normal working hours.

Frequency:

- 0 Weekly
- 0 Quarterly
- 0 Annually
- 0 Other ______
- (x) Monthly
- 0 Semi-Annually

Coverage:

- 0 Scheduled Maintenance Only
- (x) Scheduled+Unscheduled Maintenance
- 0 Repair Materials
- (x) 24-Hour Response

Equipment Covered: See equipment schedule H

**COMPRESSOR ASSEMBLY**

Air Cooled - Startup Inspection

1. Review manufacturer's recommendation for startup.
2. Energize crank case heater per manufacturer's recommendation for warmup.
3. Remove all debris from within and around unit.
4. Visually inspect for leaks.
5. Check belts, pulleys and mounts. Replace and adjust as required.
6. Lubricate fan and motor bearings per manufacturer's recommendation.
7. Inspect electrical connections, contactors, relays and operating/safety controls
8. Check motor operating conditions
9. Check and clean fan blades as required.
10. Check and clean coil. Straighten fins as required.
11. Check vibration eliminators. Replace or adjust as required.
12. Check compressor oil level, acid test oil and meg hermetic motor. Change oil and refrigerant filter drier as required.
13. Check and test all operating and safety controls
14. Check operating conditions. Adjust as required.

Mid-Season Inspection

1. Visually inspect for leaks.
2. Lubricate fan bearings per manufacturer's recommendation.
3. Lubricate motor bearings per manufacturer's recommendation.
4. Check belts and sheaves. Replace and adjust as required.
5. Clean and straighten fins as required.
6. Check operating conditions. Adjust as required.

Air Cooled Seasonal Shutdown

1. Review manufacturer's recommendation for shutdown.
2. Verify economizer operation (if applicable).
3. Visual inspection of leaks.
4. Verify auxiliary heater operation (if applicable).
5. Inspect belts and filters.
6. Test all safety controls

Optional to Above Procedures

() 1. Clean coil with chemical and high pressure spray. (See Water Treatment)

**Figure 2.2** *(Continued)*

employees training, professional development programs, and career paths that are not possible in-house. Therefore they are more likely to be better able to attract and retain the better employees.

The facility maintenance organization only needs to pay for the manpower and services it requires. In-house personnel are more likely than vendors to be distracted by other jobs or influenced by organizational polities and tradition. The productivity of in-house personnel is often diminished

**Water Cooled - Startup Inspection**

1. Review manufacturer's recommendation for startup.
2. Energize crank case heater per manufacturer's recommendation for warmup.
3. Visually inspect for leaks.
4. Vent system of trapped air.
5. Inspect electrical connections, contactors, relays and operating/safety controls.
6. Check vibration eliminators. Replace or adjust as required.
7. Check compressor oil level, acid test oil and meg hermetic motor. Change oil and refrigerant filter drier as required.
8. Check and test all operating and safety controls
9. Check operating conditions. Adjust as required.

**Mid-Season Inspection**

1. Visually inspect for leaks.
2. Check operating conditions. Adjust as required.

**Air Cooled Seasonal Shutdown**

1. Review manufacturer's recommendation for shutdown.
2. Verify economizer operation (if applicable).
3. Visual inspection for leaks.
4. Verify auxiliary heater operation (if applicable).
5. Inspect belts and filters.
6. Test all safety controls.

**Optional to Above Procedure**

() 1. Inspect and punch condenser tubes.

**Fan Assembly - Annual Inspection**

1. Check and clean fan assembly.
2. Lubricate fan bearings per manufacturer's recommendations.
3. Lubricate motor bearings per manufacturer's recommendations.
4. Check belts and sheaves. (Replace and adjust as required.)
5. Tighten all nuts and bolts.
6. Check motor mounts and vibration pads. (Replace and adjust as required.)
7. Check motor operating conditions.
8. Inspect electrical connections and contactors.
9. Lubricate and adjust associated dampers and linkage.
10. Check fan operation.
11. Clean outside air intake screen.
12. Check and clean drains and drain pans.
13. Check and clean strainers, check steam traps and hand valves.
14. Check filter advancing mechanism. Lubricate and adjust as required.
15. Inspect filters.
16. Check heating and cooling coils.
17. Inspect humidifier.

**Semi-Annual Inspection**

1. Lubricate fan bearings per manufacturer's recommendations.
2. Lubricate motor bearings per manufacturer's recommendations.
3. Check belts and sheaves. (Replace and adjust as required.)
4. Clean outside air intake screen.
5. Check filter advancing mechanism. Lubricate and adjust as required.
6. Inspect filters.
7. Check heating and cooling coils.
8. Check humidifier.

**Figure 2.2** *(Continued)*

because they tend to stick with comfortable work routines and avoid change. Outsourcing reduces an organization's need and costs to maintain inventories of supplies, materials, parts, tools, and equipment. Furthermore, it is likely that due to the economy of scale a vendor may be able to purchase these items at substantially lower costs than can otherwise be done in-house.

SCHEDULE E

AIR FILTER SERVICE

Siebe will furnish and install replacement media for the following air handling units:

**SEE SCHEDULE H FOR EQUIPMENT COVERED**

| A. H. Units | # Filters | Size | Changes/Year | Type Filter |
|---|---|---|---|---|
| | | | | |
| | | | | |
| | | | | |

***AIR FILTERS ON CONVECTORS TO BE CHANGED ONCE PER YEAR.

***AIR FILTERS ON ALL OTHER UNITS TO BE CHANGED QUARTERLY.

If additional changes are required, they will be made with the mutual consent of the owner in regard to cost for this additional service.

**Figure 2.2** *(Continued)*

When considering outsourcing, organizations should not overlook the possible disadvantages. Since the organization must pay for the profit and overhead of the vendor's employees, the vendor's hourly cost to the organization may be greater than in-house costs. Outsourcing removes the organization's direct control over the individual who actually perform the work. The goals and priorities of the vendor may not be the same as those of the organization. On-site supervision provided by vendors may change more frequently than that provided in-house. Outsourcing will

SCHEDULE G

---

**ADDITIONAL SERVICES**

In addition to the services listed in Schedules A through F, Siebe will furnish the additional services and/or special routines as listed below:

**REPLACEMENT OF WORN PARTS OR COMPONENTS:** ( From Schedules A&B )

Replacement of worn parts or components identified as being part of the mechanical equipment listed on Schedule H is not included in this agreement, with the exception of refrigerant for the air conditioning system.

Refrigerant will be provided when necessary at no additional cost. Robertshaw also agrees to make required repairs and replacements to the equipment listed on Schedule H with the owner's authorization and will bill the owner for the parts and components at the prevailing material rates.

Replacement of major equipment and/or components such as AC units, boilers, compressors, etc., are not included. In such instances, Robertshaw will either provide a proposal to do the work or will make the replacement with the owner's authorization and bill the owner for the parts and labor at the prevailing rates.

**Figure 2.2** *(Continued)*

require the organization to expend allocation resources to procure and manage the services which are outsourced.

The outsourcing processing often causes an organization to evaluate the cost/benefit for these services and realistically assess their actual needs. In addition, the facility has the ability to use the vendor's expertise as a consultant.

Outsourcing is an important part of many flexible staffing strategies. According to Professor Michael Bur of the Harvard Business School, "Expertise and excellence come from specialization."

Despite the potential benefits, Facility Managers are often concerned about the potential loss of control over services that are outsourced. By hiring outsiders to perform important functions facility managers must assume additional responsibilities. These responsibilities include the development of specifications, solicitation of services, preparation and award of the contract, and control of the

# Master Maintenance Agreement

**AGREEMENT FOR**
**MASTER MAINTENANCE SERVICE**

TO: ______________________ (Purchaser - herein called You)

BUILDING LOCATION ______________________

11510 Falls Road — 11510 Falls Road

Potomac, Maryland 20854 — Potomac, Maryland

Company (herein called We) will provide MASTER MAINTENANCE SERVICE on the elevator equipment in the above building and described below (herein called the equipment) on the terms and conditions set forth herein.

| No. Elevators and Type | Manufacturer | Serial No. |
|---|---|---|
| One Hydraulic Passenger | | E-90263 |

EXTENT OF COVERAGE

We will:

Regularly and systematically examine, adjust, lubricate and, whenever required by the wear and tear of normal elevator usage, repair or replace the equipment (except for the items stated hereafter), using trained personnel directly employed and supervised by us to maintain the equipment in proper operating condition.

Furnish all parts, tools, equipment, lubricants, cleaning compounds and cleaning equipment.

Relamp all signals as required during regular examinations only.

Periodically examine and test the hydraulic system and/or governor, safeties and buffers on the equipment, at our expense, as outlined in the American National Standard Safety Code For Elevators and Escalators, A.N.S.I. A17.1, current edition as of the date this agreement is submitted. It is expressly understood and agreed that we will not be liable for any damage to the building structure occasioned by these tests.

ITEMS NOT COVERED

We assume no responsibility for the following items, which are not included in this agreement:

The cleaning, refinishing, repair or replacement of

- Any component of the car enclosure including removable panels, door panels, sills, car gates, plenum chambers, hung ceilings, light diffusers, light fixtures, tubes and bulbs, handrails, mirrors, car flooring and floor covering.
- Hoistway enclosure, hoistway gates, door panels, frames and sills.
- Cover plates for signal fixtures and operating stations.
- Intercommunication systems used in conjunction with the equipment.
- Main line power switches, breakers and feeders to controller.
- Emergency power plant and associated contactors.
- Emergency car light and all batteries, including those for emergency lowering.
- Smoke and fire sensors and related control equipment not specifically a part of the elevator controls.
- Jack unit cylinder, buried piping and buried conduit.

**Figure 2.3** Maintenance master agreement.

contracted work while in process and on completion. In addition, the providers' performance must be measured and evaluated. Clearly outsourcing is not a simple solution.

The preferred way to procure outsourced services is by competitive solicitation process. This process required written solicitation package, a vendor list, and bid evaluations. The solicitation package should include a detailed written

PRORATED ITEMS

The items listed on the schedule below show wear and will have to be replaced in the future. To provide you with the maximum of service from these items, we are accepting them in their present condition with the understanding that you agree to pay, in addition to the base amount of this agreement, an extra at the time the items listed are first replaced by us. Your cost for the replacements will be determined by prorating the total charge of replacing the individual items. You agree to pay for that portion of the life of the items used prior to the date of this agreement, and we agree to pay for that portion used since the date of this agreement.

SCHEDULE OF PARTS TO BE PRORATED

| NAME OF PART | DATE INSTALLED |
|---|---|
| -NONE- | |

HOURS OF SERVICE

We will perform all work hereunder during regular working hours of our regular working days, unless otherwise specified. We include emergency minor adjustment callback service during regular working hours of our regular working days.

If overtime work is not included and we are requested by you to perform work outside of our regular working hours, you agree to pay us for the difference between regular and overtime labor at our regular billing rates.

PURCHASER'S RESPONSIBILITIES

- Possession or control of the equipment shall remain exclusively yours as owner, lessee, possessor or custodian.
- Your responsibility includes, but is not limited to, instructing or warning passengers in the proper use of the equipment, taking the equipment out of service when it becomes unsafe or operates in a manner that might cause injury to a user, promptly reporting to us any accidents or any condition which may need attention and maintaining surveillance of the equipment for such purposes.
- You will provide us unrestricted access to the equipment, and a safe workplace for our employees.
- You will keep the pits and machine rooms clear and free of water and trash and not permit them to be used for storage.
- You agree that you will not permit others to make changes, adjustments, additions, repairs or replacements to the equipment.

TERM

This agreement is effective as of ___August 28___, 19_88_ (the anniversary date) and will continue thereafter until terminated as provided herein. Either party may terminate this agreement at the end of the first five years or at the end of any subsequent five-year period by giving the other party at least ninety (90) days prior written notice.

This agreement may not be assigned without our prior consent in writing.

**Figure 2.3** *(Continued)*

Statement of Work. "Statement of Work" means the "specifications" used to describe the requirements of the services. The form of the Statement of Work may be either term or completion. A *term* form expresses a level-of-effort requirement during a period of time. A *completion* form describes providing a completed product or service.

It is important to write clear statements of work for the services that are to be outsourced. The courts usually inter-

CONDITIONS OF SERVICE

No work, service or liability on the part of , other than that specifically mentioned herein, is included or intended.

The parties hereto recognize that with the passage of time, equipment technology and designs will change. We shall not be required to install new attachments or improve the equipment or operation from those conditions existing as of the effective date of this agreement. We have the responsibility to make only those adjustments, repairs or replacements required under this agreement which are due to ordinary wear and tear and are disclosed to be reasonably necessary by our examination. You agree to accept our judgement as to the means and methods to be used for any corrective work. We shall not be required to make adjustments, repairs or replacements necessitated by any other cause including but not limited to, obsolescence, accidents, vandalism, negligence or misuse of the equipment. If adjustments, repairs, or replacements are required due to such causes, you agree to pay us as an extra to this agreement for such work at our regular billing rates.

We shall not be required to make tests other than those specified in the extent of coverage, nor to install new attachments or devices whether or not recommended or directed by insurance companies or by federal, state, municipal or other authorities, to make changes or modifications in design, or make any replacement with parts of a different design or to perform any other work not specifically covered in this agreement.

It is understood, in consideration of our performance of the service enumerated herein at the price stated, that nothing in this agreement shall be construed to mean that we assume any liability on account of accidents to persons or property except those directly due to negligent acts of Dover Elevator Company or its employees, and that your own responsibility for accidents to persons or properties while riding on or being on or about the aforesaid equipment referred to, is in no way affected by this agreement.

We shall not be held responsible or liable for any loss, damage, detention, or delay resulting from causes beyond our reasonable control, including but not limited to accidents, fire, flood, acts of civil or military authorities, insurrection or riot, labor troubles, including any strike or lockout which interferes with the performance of work at the building site or our ability to obtain parts or equipment used in the performance of this agreement. In the event of delay due to any such cause, our performance under this agreement will be postponed without liability to us by such length of time as may be reasonably necessary to compensate for the delay. In no event will we be responsible for special, indirect, incidental or consequential damages.

PRICE

The price for the service as stated herein shall be One Hundred and Fifty Dollars ($ 150.00 ) per month, payable monthly in advance upon presentation of invoice. You shall pay as an addition to the price, the amount of any sales, use, excise or any other taxes which may now or hereafter be applicable to the services to be performed under this agreement.

This price shall be adjusted annually and such adjusted price shall become effective as of each anniversary date of the agreement, based on the percentage of change in the straight time hourly labor cost for elevator examiners in the locality where the equipment is to be examined. For purposes of this agreement, "straight time hourly labor cost" shall mean the straight time hourly rate paid to elevator examiners plus fringe benefits which include, but are not limited to, pensions, vacations, paid holidays, group life insurance, sickness and accident insurance, and hospitalization insurance. The straight time hourly labor cost applicable to this agreement is $ 24.035 of which $ 7.35 constitutes fringe benefits.

A service charge of 1½% per month, or the highest legal rate, whichever is less, shall apply to delinquent accounts. In the event of any default of the payment provisions herein, you agree to pay, in addition to any defaulted amount, all our attorney fees, collection costs or court costs in connection therewith.

**Figure 2.3** *(Continued)*

pret ambiguities in a specification against the drafter. A good specification defines the minimum requirements, not the personal preferences of the users. The specification should also include provisions to establish that the supplier can meet all the user requirements. Some work like landscape maintenance, elevator maintenance, roof repairs, and custodial services are easily definable, and a very definitive "do it this way" statement of work is appropriate. However, other work such as architectural, engineering, and consulting services are more abstract and difficult to define. In these situations, a performance or "this is what we think we

SPECIAL CONDITIONS

ADDITIONAL PROVISIONS

This instrument contains the entire agreement between the parties hereto and is submitted for acceptance within 30 days from the date executed by us, after which time it is subject to change. All prior negotiations or representations, whether written or verbal, not incorporated herein are superseded. No changes in or additions to this agreement will be recognized unless made in writing and signed by both parties.

No agent or employee shall have the authority to waive or modify any of the terms of this agreement.

We reserve the right to terminate this agreement at any time by notice in writing should payments not be made in accordance with the terms herein.

Should your acceptance be in the form of a purchase order or similar document, the provisions, terms and conditions of this agreement will govern in the event of conflict.

ACCEPTANCE BY YOU AND SUBSEQUENT APPROVAL BY AN EXECUTIVE OFFICER OF DOVER ELEVATOR COMPANY WILL BE REQUIRED BEFORE THIS AGREEMENT BECOMES EFFECTIVE.

Accepted: ____________________
(Full Legal Company Name or Individual Purchaser)

By: Mark Aulenbe[signature]
(Signature of Authorized Official)

____________________
(Type or Print Name)

Title Executive Director
(Type or Print)

Date Signed: September 23, 1988

BILLING ADDRESS:

By: Charles L. Stump Jr. [signature]
Chas Stump

Date Signed: ____________________

APPROVED:

By: ____________________

Title: ____________________

Date Signed: ____________________

Figure 2.3 *(Continued)*

want to accomplish" statement of work is usually most appropriate.

When dealing with large complex issues such as facility planning or construction, it is usually beneficial to talk with prospective vendors prior to preparing the Statement of Work to learn more about the range of the services they offer. It is also helpful to break larger projects into smaller more manageable phases each with their own Statement of Work.

A detailed description or Statement of Work is the specifications used to describe the requirements for contract services. There are two types of specifications: *term* and *completion*. A *term statement of work* describes a level-of-effort for a period of time, for example, the requirement to provide a boiler operator for the month of December. A *completion type statement of work* describes the actual work the user requires. The contractor must provide all the resources (i.e., manpower, supervision, supplies, tools, and equipment) to accomplish the work. A poorly written statement of work can result in unreasonable prices, failure to obtain offers, failure to obtain the denied level of effort from the contractor, etc. Words that are vague such as "reasonable," "clean," "functional," "and/or," etc., should not be used. Words such as "including" and "similar" and phrases such as "good workmanship" and "neatly finished" should be avoided.

The user must identify the services that are desired to be procured. These requirements must be described in a written document called a "specification." The user then must determine what type of contract and method of procurement of these services.

There are basically two methods of procurement: *sealed bidding* (*IFB*) and the *negotiated method* (*RFP*). Sealed bidding is generally the preferred method of contracting. This method generally affords the maximum level of competition and thus results in the lowest prices. The main goal of sealed bidding is to give all qualified sources an opportunity to bid competitively on an equal and fair basis. Under this process the offeror must solicit bids (i.e., prices or price-related factors responsive to the solicitation). This type of solicitation is usually called an *invitation for bids* (*IFB*).

An Offeror's submission is "nonresponsive" if it does not comply with the terms of the solicitation in a manner which could affect price, quantity, quality, and/or delivery. Generally the bids are opened publicly and remain effective for a set period of time. Offerors may not change or withdraw their bids after they are opened except under certain conditions. IFB are intended to be awarded to the Offeror who submitted the lowest responsive offer. In this process offers must submit bids that are responsive to the solicitation. However, for oral bidding to be viable the following conditions must exist:

- Sufficient time to complete the process
- Adequate competition is expected
- Satisfactory specifications
- Price and price-related factors are sufficient for determining the award

If the above conditions are not met, the negotiated method of procurement should be pursued. Unlike the sealed bidding, the negotiated methods proposals do not have to be entirely responsive to the solicitation, and they may or may not be opened in public.

These proposals may be withdrawn at any time before an award is made. This type of solicitation is usually called a *request for proposals* (*RFP*).

The goal of negotiation of competition. Negotiation allows greater flexibility through the use of discussion and the opportunity for prospective contractors to modify their offers. If discussions are held with offerors, they are usually encouraged to submit the "best and final" offer.

The award of RFP is generally more time consuming and complicated than an IFB. RFPs should be evaluated by a rational set of predetermined technical and cost evaluation criteria. The solicitation should give the offerors reasonable notice of the relative importance of the solicitation criteria. However, it is not necessary to identify the weights of each criteria in the solicitation. Technical criteria can be of the threshold type. These are mandatory requirements which when applied will result in either a "yes" or "no" response from the prospective offeror such as, "Does the offeror have at least five years of relevant experience?". The other type of criteria is in "variable" criteria. This type of criteria is scored according to the degree it fulfills this requirement. An example of a question that should be evaluated by the criteria is the management plan the contractor proposes to implement, or the key personnel that the offeror proposes to use, to fulfill the requirements of the specification. Frequently the evaluation of technical proposals is performed by a group or panel, rather than a single individual.

On occasion it may be necessary to procure services through other than a competitive process in order to:

- Maintain continuity of services
- Maintain existing business partnership
- Acquire contractors with unique qualifications/experience

Other procurement methods may be developed using various elements of the sealed bid and negotiation.

Communication of the user's requirements to prospective vendors can be facilitated by using a uniform contract and specification format. A conference and/or tour of the job site should be scheduled prior to the recruitment of bids or proposals. A conference should be scheduled to provide prospective bidders an opportunity to discuss and clarify potential discrepancies. A job site visit allows prospective offerors opportunity to observe conditions which may hinder, or facilitate, the ability to fulfill the requirements of the specification. As soon as the solicitation process is completed, a contact needs to be executed and administered with the successful vendor. In addition, the unsuccessful vendors need to be notified of the outcome of the solicitation.

When outsourcing is employed, in the facility management organization, there is a need to designate someone in-house to be responsible to administer development of specifications, the solicitation and award process, and inspect the work of the contractor, as well as to administer and ensure compliance of the contract after it is awarded.

In small organizations, a single individual may be responsible to fulfill all these duties. Consultants may be used as the circumstances warrant to develop the specification. It is prudent to use a lawyer experienced in contract law and facility operations and maintenance issues to develop the required contract documents. For example, the contract should include provisions which protect the interests of the facility management organization. Such provisions include, but are not limited to, the ability of the facility management organization to amend, extend, or terminate the contract for the convenience of the facility management organization. In addition, the facility management should designate an individual with assigned specific authority and responsibility to enter into and administer contracts for outsourced services on its behalf. The various duties necessary to administer contracts include:

- Work initiation conference;
- Schedule and coordinate the contractors activities;
- Implement an effective quality assurance program;
- Ensure compliance will all the terms and conditions of the contract specification, terms, and conditions;
- Verify and approve invoices for payment;
- Pay invoices in a timely manner for work that the contractor completed pursuant to the terms of the contract;
- Maintain an accountability for all work performed and costs that are incurred; and
- Ensure work is performed within budget allocation.

Service contracts may be amended to reduce or increase the scope of work during contract performance.

# NOTES

# NOTES

Chapter

# 3

# Operations and Maintenance Plans

## Management Operational Plans

### Work control method and procedures

The facility manager's primary goal is to manage resources wisely by providing responsive, high-quality maintenance and repair services to all entities being supported. To accomplish this mission the facility manager must establish well-defined procedures and the organizational structure to fulfill this work. The procedures involve coordination and planning to ensure that all the elements of skills, tools, equipment, and materials are synchronized at the right time and in the right mix to produce the desired result of satisfying the customer while concurrently controlling costs. The organization must be such that all components are orchestrated to function as a smoothly running team. Work control methods allow work requirements to be identified, screened and evaluated, planned and scheduled, checked and inspected, closed out and cost accounted, results recorded, analyzed, and measured, and finally, feedback given to the customer. A typical work control cycle is shown in Fig. 3.1. Finally, it is important to point out

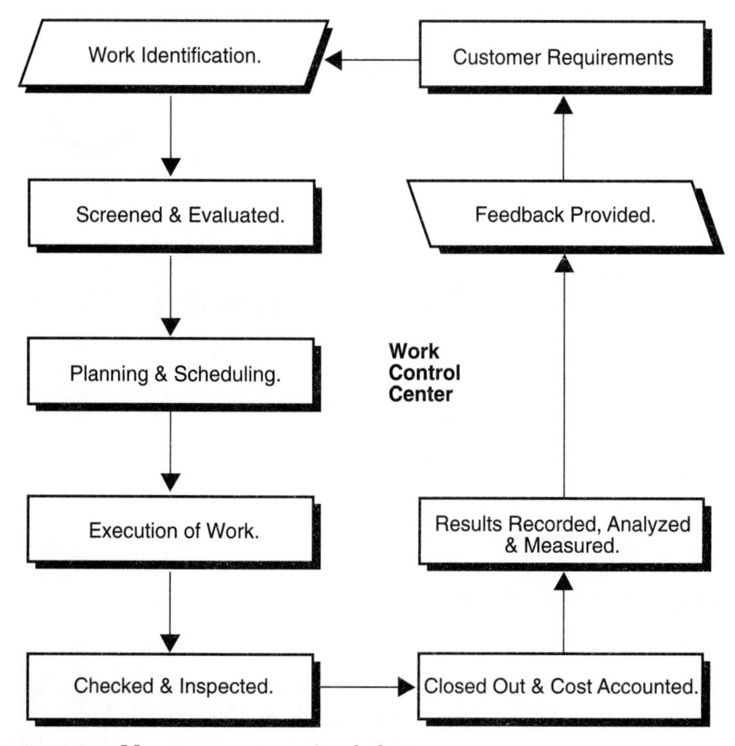


**Figure 3.1** Management operational plan.

that there is no standard model to follow. One method would be to establish a work control center capable of controlling all work received in the facility department. Work control centers can range in size from a one-person operation, as an added duty, to an entire department supporting a large municipality or billion dollar corporation.

## Work control center

The work control center is the "heartbeat" of any facilities organization. This is the central point where all work requirements are funneled, then coordinated, planned, costed, scheduled, and measured. As the primary interface between the customer and the organization the work control center can significantly influence the facility management

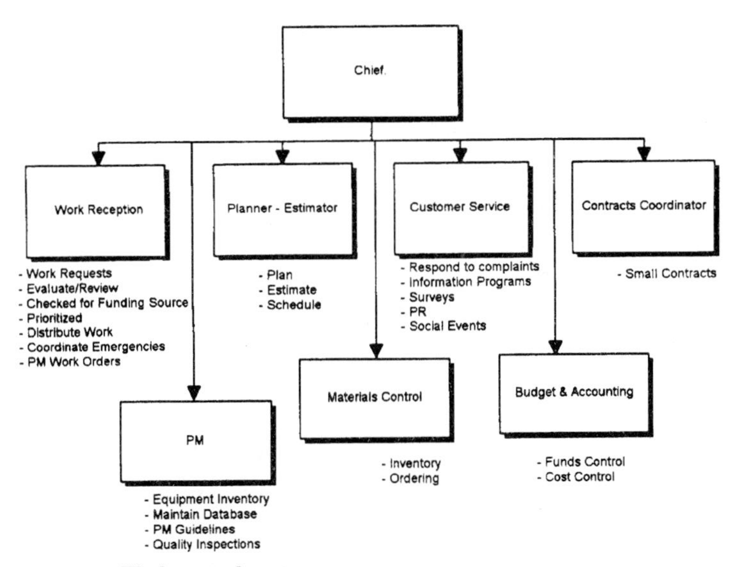


**Figure 3.2** Work control center.

image. Continuous coordination with workers and "closing the loop" or providing feedback to customers is essential to develop the professional reputation of the department both internally within the facility management department and externally with customers. Figure 3.2 depicts a large work control center.

### Work control center responsibilities

- Centralize communications. One central telephone number for customers to call. The work control center manages the department communications systems.
- Receive and record all work requests.
- Operate a work management control system either manually, or computerized, using a computerized maintenance management system (CMMS).
- Coordinate all work performed with both workers and customers.

- Coordinate and assign work priorities.
- Coordinate small contract work.
- Develop work order estimates.
- Plan and schedule work.
- Operate preventive and predictive maintenance programs and monitor quality control inspections.
- Assess productivity.
- Manage emergencies and special support work such as snow removal, utility outages, and special events.
- Track work backlog.
- Respond to customer complaints.
- Control supplies, parts, and materials inventory.
- Operate cost accounting and budgeting systems.
- Market "best practices" and institute continuous work improvement procedures.
- Measure performance of work against established goals and standards (both quantity and quality).

### Staffing and location

The work control center should be accessible and convenient to all departmental components. Each of its organizational components should have access to data and information such as historical facility information, O&M manuals, "as built" drawings, and warranties for equipment and systems.

Ideally, the work control center area should be bright and open and located in a central department location. It should be the main hub of communications for the department and should include:

- One central telephone number with multiple work stations. Sufficient work receptionists to respond to customer requests and key those requests into the CMMS.
- Access to read the energy management control system screens in order to triage "too hot" or "too cold" calls.

- Radio communications with mechanics and technicians in the field in the event of emergency calls.
- Facility and utility maps in order to pinpoint facility and utility emergency locations and for special support requirements. Have on-hand and accessible current policies, procedures, and operations plans.
- Access to specialized contractor telephone numbers, on a 24-hour basis, in the event of emergencies.
- Personnel who have been trained in customer service techniques, are polite, and are willing to assist the person or group being served.

### Organizational elements

**Work reception.** Work requirements are generated from various means: customers, either verbally or through the completion and submission of a work request form; and operations and maintenance personnel as a result of preventive and predictive maintenance tasks, facility tours, and facility inspection programs. These requirements are received telephonically, from walk-in customers, in writing, and in some cases through electronic mail. Upon receipt, work requests are evaluated for validity, reviewed for compliance with regulatory requirements and feasibility, checked for funding source, prioritized, assigned a project number, and entered into the CMMS. If additional information is required, the caller or initiator of the request is contacted and requested to provide additional detail. Service work requests are then generated and forwarded (electronically or manually) to the appropriate shop, or zone, for accomplishment. In the event of emergencies, calls are made via radio or other expeditious means to field mechanics.

Work reception centers vary in size and complexity depending on the size of the facility department organization. In small companies, customers can contact the "maintenance man" or, for example, if in a hotel, the front desk, or the front office secretary in small companies. In larger organizations such as large companies, government agencies, or large universities, users would contact the customer service center or "trouble desk." The point is that regardless of size, every facilities organization should have a formal method for customers to request assistance.

**Planner-estimator.** The normal shortage of facility management resources dictates that work be carefully planned and estimated to ensure efficient accomplishment. The planner-estimator must be an experienced and highly competent individual who can generate work that is easily understood by facility workers. This individual is responsible for planning and estimating in-house work which could involve only one trade to more complex projects involving multiple trades and requirements. To ensure accurate cost estimating, realistic scheduling, and timely materials coordination, it is important that the work plan be accurate, understandable, and thorough. Sketches and drawings should be provided describing technical characteristics, sizes, and dimensions of the job. Sources of information for use in planning, scheduling, and measuring the efficiency of work performance should include engineered performance standards (EPS) from recognized estimating handbooks and computer software such as F.W. Dodge and R.S. Means. A bill of materials specifying quantities of materials and items of equipment required for the job should also developed and provided to cognizant supervisors, and the materials coordinator, for work planning and performance tasks.

**Materials control.** The materials control section stores materials, parts, and supplies to support planned operations and maintenance requirements, preventive maintenance tasks, and other repair and new work activities. Its main purpose is to provide timely parts, supplies, and materials while concurrently controlling the availability of the inventory which ultimately impacts the organization's budget. Stated in simpler terms: to provide materials, parts, and supplies at the lowest possible cost to the right place and at the right time. In many facility departments space for storage or warehousing of parts, supplies, and materials is insufficient. One alternative is to partner with a local supplier who is willing to store materials, parts, and supplies, at no extra cost, and guarantee 24-hour delivery, in return for preferential treatment of purchases. Large-volume items such as filters and light bulbs lend themselves very nicely to this cooperative type of agreement. In addition to providing storage of parts and supplies, most suppliers will also give discounts of up to 20 percent.

**Budget and accounting.** The work control center should be responsible for developing the budget, and performing cost accounting and fund control for the facility department. In large facility organizations these functions are accomplished by the budget and accounting section. This section is responsible for planning, supervising, and coordinating the preparation and analysis of the facility management budget. Once the budget has been formulated and approved, the facility manager is responsible for its proper execution. The execution phase is then divided into two categories: (1) funds control and (2) cost control.

- *Funds control.* The status of various facility management accounts must be closely monitored to ensure that the budget is not over obligated. One method that is commonly used is a monthly review of the operating budget to analyze and compare it to the year-to-date actual expenses and encumbrances. This is then compared to the overall budget and gives the facility manager a real-world view of expenditures against budgeted funds.
- *Cost control.* This requires good tracking and record keeping. Without it the facility manager will not have good job control. A good cost control system provides for tracking by budget item, work classification, and job function. While not totally exact, it will give the facility manager a good approximation of cost.

**Preventive maintenance (PM).** The preventive maintenance manager operates the PM program for the department. This can be done manually or with a computerized maintenance management system (CMMS). The latter is necessary in order to manage large electrical and mechanical systems consisting of thousands of items. A key component of a good PM program is being able to recognize which major equipment components, units, and other items should be included. Consideration must be given to problems that could result in the future that require further evaluation and examination. The PM section should also maintain cognizant records, publications, and specialized tools to help with diagnosing and predicting potential equipment and systems defects. Records include PM inspection lists and schedules, access to "as built" drawings, either paper copies or through the computer assisted digital data (CADD) system, and repair histories of equipment and components (organizations that are computerized can have this information located in their CMMS). Publications include parts, service, and operations manuals, and other engineering data. Specialized tools common to predictive maintenance include infrared imaging cameras, vibration analysis collectors, bore scopes, temperature measuring devices, ultrasonic testing devices, listening devices such as a stethoscope, and other high dollar equipment. All of the above can be used by various entities throughout the facility department when it is necessary to determine the repair history of a system or component, repair parts reference, asset value, and for planning and scheduling information.

**Customer service.** While everyone in the facility department must be customer service oriented, it is beneficial to have a section dedicated to this function and also focused on improving the image of the department, and marketing its positive contributions. Excellent customer service is a prerequisite for success. It exists when the facility department meets and exceeds customer expectations for service. In order for this to happen, the department must have a program in place which allows it first to provide excellent service and second to find ways of reminding the customer of the great service it does give. What customers perceive is reality to them. This work control center section, then, has the goal of shaping perceptions.

**Work identification.** Work for the facility department is generated when a customer identifies a requirement and asks that it be accomplished. Requests can come from externally supported customers or facility staff personnel through the normal discharge of their duties. Facility personnel generate routine work requests, for example, in the performance of preventive maintenance, through scheduled inspections such as monthly fire extinguisher inspections, and by simply walking and looking at their facilities on a daily basis. Normally, changes occur in the physical condition of the facility from age, environment, and use; by changes in regulatory requirements; and by requirements for construction, including alteration, due to expansion and changes in mission and operational needs.

The work receptionist, located in the work control center, is the primary point of contact for all customers. In a facility department the work receptionist has the responsibility for receiving all requests for work and entering them into the work management and control system. A formal method must be in place for receiving these work requests. The work request form is the management tool established for just this reason. It is an official document used to record a description of the work to be done, who requested it (and how to contact), when it must be completed, a budget number if the work is reimbursable, and finally, the signature of an approving individual who is authorized to obligate funds for that particular department. Work requests can be generated in several different ways: FAX, telephone, mail, and walk-up. Figure 3.3 depicts a sample written work request form and Fig. 3.4 illustrates an electronic work request form. The initiator of the work request normally completes the form. If the work is called in by telephone, the work receptionist will complete the form. Written work request forms normally consist of several carbon (manifold) sheets used by various cognizant personnel to track the work from inception through completion. One of the manifold copies is provided customers for their record.

## Facility Work Request

Section - 1

| Control Number | Time: | Date: ____ / ____ / ____ |
|---|---|---|
| Title | | |

Section - 2

| Facility Name: | Location: |
|---|---|
| Description of Work: | |
| Justification: | |
| Request for Cost Estimate Only ? Yes ☐ No ☐ | Sketch Attached? Yes ☐ No ☐ |
| Requested Completion Date: ____ / ____ / ____ | |
| Requesting Organization: | Cost Center: |
| Requestor Contact: | Telephone: |
| Requestor Authorized Signature:<br>Name: | Date: ____ / ____ / ____ |

Section - 3

| Type of Work: | Classification: | |
|---|---|---|
| Cost Estimate: | $ | Date: ____ / ____ / ____ |

| Funding: | Cost Center | Amount | Percentage |
|---|---|---|---|
| Source 1 | __________ | $ | __________ % |
| Source 2 | __________ | $ | __________ % |

| Division of Facilities Authorization: | Date: ____ / ____ / ____ |
|---|---|
| Division of Facilities Contact: | Telephone: |
| Division of Facilities Comments: | |

**Figure 3.3** Sample written work request.

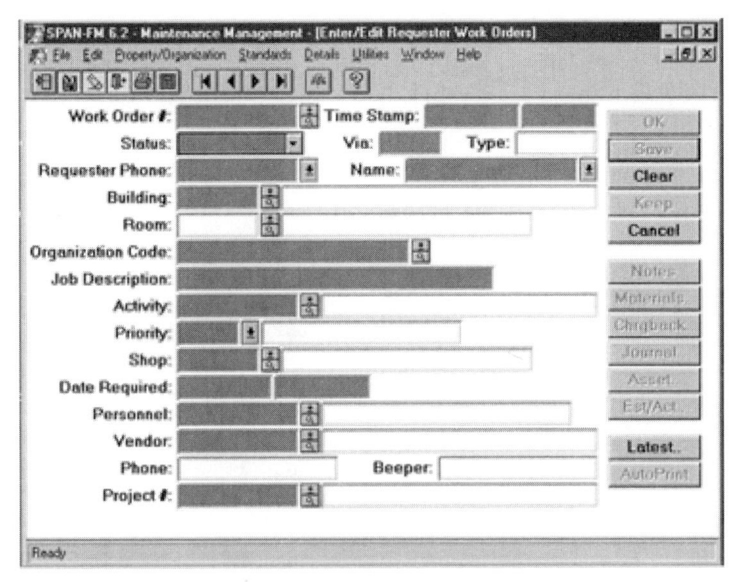


**Figure 3.4** Electronic work request form. (Copyright © 1998, Innovative Tech Systems, Inc.)

Once the work request is received and entered into the work management and control cycle, it becomes a generic work order that is assigned a work order number. The work control center then tracks each work order as to status: waiting scheduling, waiting materials, ongoing, or completed. A typical work order flow is shown in Fig. 3.5.

**Work screening and evaluation.** The work receptionist then classifies each work request as either maintenance, repair, or new work (see Fig 3.6).

Maintenance work is defined as work performed to keep the facility operating and prevent equipment and systems breakdowns. Repair work is necessary to fix something that has already failed. Finally, new work is work that is being added to expand, enhance, or reconfigure a facility.

Once work has been classified, it is categorized into one of four categories: (1) service orders, (2) work orders, (3) standing operating orders, and (4) preventive maintenance work orders.

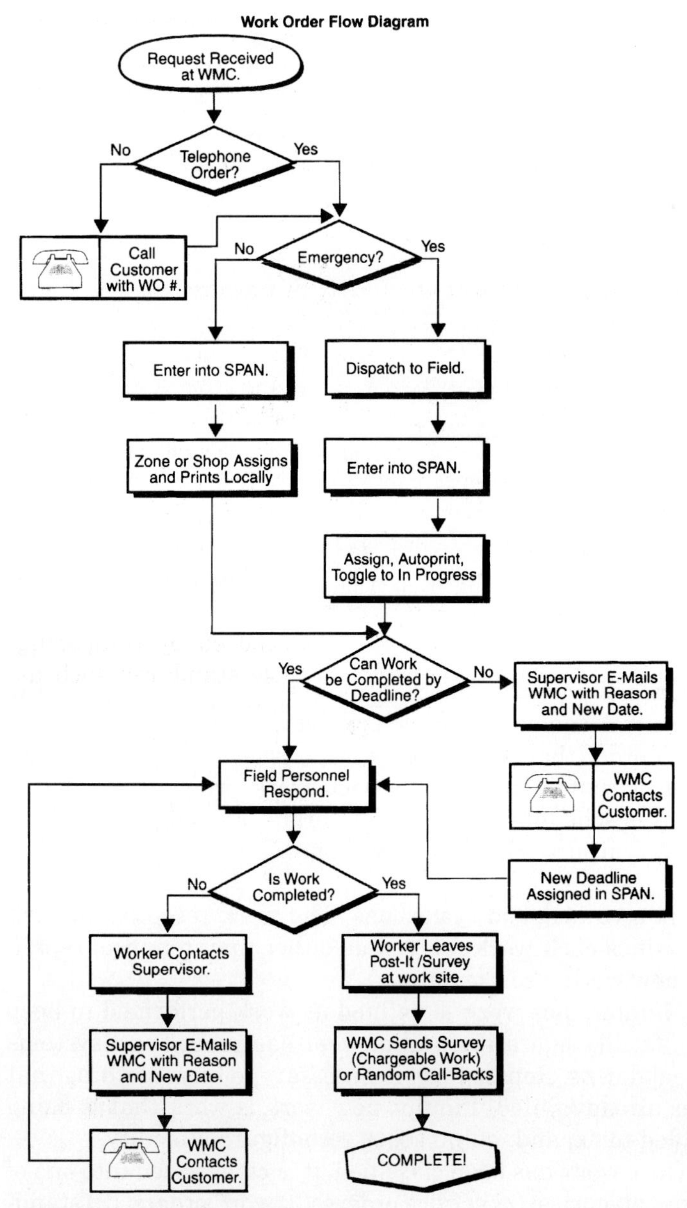


**Figure 3.5** Work order flow diagram.

| | |
|---|---|
| Maintenance: | Painting, filter changes, pothole repairs |
| Repair: | Repair of plumbing leaks, broken glass, torn carpet |
| New work: | Installing new electrical outlet, constructing a new addition to a building |

**Figure 3.6** Examples of work classifications.

## Service orders (SO)

Service orders are small, service-type maintenance jobs that require immediate attention and cannot be deferred. Normally, there are time and cost limitations established for this type of work, for example, $500 and 8 worker-hours of effort. Planning, estimating, and scheduling are not performed. Typical service orders include leaking roof, minor electrical and plumbing work, and broken window repairs. There should be continuous review and analysis of effort expended on service orders. This should be done by shop and/or by zone, depending on how the department is organized, and by similar jobs to determine workload and efficiency. Productivity can also be measured by comparing work effort to established performance standards such as EPS, F.W. Dodge, and R.S. Means mentioned earlier.

The number of service orders can be reduced by strictly enforcing work priorities, forcing customers to pay for damages caused as a result of vandalism or for repairs other than fair wear and tear, adhering to established performance standards, and applying strong preventive and predictive maintenance programs.

## Work orders (WO)

Work orders can include maintenance, repair, and new work. They are formally planned, estimated, scheduled, and cost accounted. This type of work can be accomplished by the in-house work force or by contract, depending on limitations that have been established. Work performed in-house is planned, estimated, and scheduled by the planner-estimator. Work performed by contract is scheduled by the contracts coordinator. In either case, upon completion of the job, completion documents showing cognizant costs and usage data are returned to the budget and accounts section for appropriate action.

### Standing operating orders (SOO)

Standing operating orders are special work orders where the specific work and the labor hourly requirements are relatively constant and predictable. They include operations, services, and routine maintenance where requirements are repetitive and can be planned annually. Examples include fire protection inspections, custodial services, utility plant operations, refuse collection and disposal, grass cutting, etc.

Work is planned, estimated, and scheduled prior to the start of the fiscal year. Careful analysis should be made of all standing operations. Scheduling must be complete and detailed. There should be periodic inspections and ongoing work sampling for each standing operation to determine the quality of the provided service and quantity.

### Preventive maintenance (PM)

Preventive maintenance is routine, recurring work such as regular servicing of HVAC systems and equipment. PM reduces emergency responses, allows work to be efficiently scheduled, and ensures that replacement parts are available when needed. PM work orders are generated in a similar manner as normal work orders.

### Work priority system

After the work is classified and categorized, it should be prioritized for work performance scheduling. Each facility department should establish its own work priority system. Normally, there are three standard levels of work priorities: (1) emergency, (2) urgent, and (3) routine. These priorities should be carefully and clearly defined and distributed to all customers for their understanding. By doing this, the job of the work receptionist should be made easier and become a little more pleasant.

**Emergency priority.** This work request designation takes priority over all other work. It requires immediate action. Work continues until the emergency is corrected. Normal response time is within 1 hour. Examples include lock-ins, gas leaks, broken water and steam lines, utility failures, hazardous and toxic spills, elevator trap calls, and stopped-up commodes (when only one is available).

**Urgent priority.** This priority involves correcting a condition that could become an emergency. Normal response times range from immediate up to 72 hours depending on availability of the workforce. Once started, the work should continue until completed dependent on availability of materials and parts. Examples include garbage disposal problems, lights that are out, pest control problems, lock changes, and appliance (refrigerator, range) problems.

**Routine priority.** This is work that is an inconvenience to the customer. The completion of this work is normally accomplished within seven calendar days. Examples include dripping faucets, screen door replacement, countertop repair, and floor replacement.

### Work approval

The work control center supervisor is responsible for work approval. This is normally delegated to the work receptionists. Experienced work receptionists can usually estimate the cost of service orders work fairly accurately, and then, using established work performance criteria as guides, assign the work to the work force. For more complex and costly work, such as work orders, the work control center supervisor should be the approval authority. This work should be examined from the standpoint of work assignment, i.e., whether it is reimbursable work or not. If reimbursable (e.g., new work), a budget number should be provided by the customer. Repetitive work, standard operating orders, and PM work, are planned on an annual basis and does not require individual work order approval.

### Planning

Once work orders are approved, work receptionists forward them to the planner-estimator. This individual then develops the detail of the scope of work; estimates the total cost of the job, which includes labor, material, length of time, and equipment required; and then prepares plans, sketches, and specifications to do the work at minimal cost and time while concurrently achieving the highest quality of work possible. To be effective at job planning the planner-estimator usually needs to query the initiator of the work request for specific detail, then visit the job site to verify information compiled to date and examine specific site conditions that will impact cost, and finally determine if permits are required. In larger, complex jobs involving multiple crafts it would be beneficial for the planner-estimator to use a planning worksheet.

### Scheduling

Scheduling is placing what has been planned into action. This is where the planner-estimator juggles the availability of trades labor hours; job site availability; special customer requirements; materials, parts, supplies, and equipment availability; weather conditions, and the relative priorities of the job. A key factor that must be considered and expected is the need for flexibility. No job runs exactly as planned and scheduled. There are always disruptions, delays, or changes that must be anticipated and planned for. One objective that every facility manager should have is to accomplish the largest portion of maintenance work on a scheduled basis. This is possible with the help of a good PM program. The facility manager will then see that as the amount of scheduled work increases, so will work force productivity. This is due to a decrease in unscheduled and emergency work.

The Association of Higher Education Facilities Managers (APPA) defines three intervals for scheduling work. These intervals are printed with permission from APPA.

**Long-term schedules.** Usually 3 months or greater, based on the PM master schedule, work order backlog, priority of the work to be done, availability of equipment and materials, time of the year (weather), and availability of trades labor, these schedules must be continuously reviewed and revised as necessary. Characteristics of a typical long-term schedule include:

- Only large jobs or projects are scheduled. Typically, large means anything with more than 24 to 32 work hours of labor. No attempt is made to schedule emergency work.
- A separate work order master schedule should be maintained for each shop so that individual skill or trade bottlenecks can be foreseen.
- Care needs to be taken to avoid exceeding the capacity of each shop. To do this, the capacity of each shop must be calculated and updated regularly.
- All available shop labor hours capacity should not be scheduled. Typically, only 50 to 80 percent of available labor hours should be scheduled. The remaining 20 to 50 percent should be reserved for absenteeism, nonproductive work, and emergencies.
- In well-run master planning scheduling systems, the work order schedule becomes frozen 1 to 2 months out; changes are allowed only in the third month and beyond. This greatly stabilizes the planning and scheduling process and contributes to higher productivity and schedule reliability. Figure 3.7 is an example of a long-term schedule.

**Weekly scheduling.** This interval of work order scheduling is for the coming week. It is developed from the long-term schedule. It should represent the facility manager's detailed commitment of resources. Characteristics of a typical weekly work order schedule include:

- While master work order schedules cover only 50 to 80 percent of shop capacity, weekly schedules plan to cover 80 to 90 percent of shop capacity.
- Jobs under 24 to 32 work hours duration are not included on the weekly schedule.

**Typical Master Schedule**

**Shop** Carpenter **Date** 3-14

| **Time Period** | **1st Half April** | **2nd Half April** | **May** | **June** | **July** | **Aug** | **Sep** | **Oct** |
|---|---|---|---|---|---|---|---|---|
| Capacity (workhours/work day) | 24 | 24 | 24 | 24 | 32 | 32 | 32 | 32 |
| No. of work days in period | 10 | 9 | 20 | 21 | 21 | 22 | 20 | 22 |
| Planned workhours available | 240 | 216 | 480 | 504 | 672 | 704 | 640 | 704 |
| Total load (workhours) | 200 | 220 | 470 | 450 | 370 | 340 | 340 | 240 |
| Net workhours available | 40 | -4 | 10 | 54 | 302 | 364 | 300 | 464 |
| **Jobs or Projects** | | | | **Scheduled Workhours** | | | | |
| # 6387 ABC Hall | 180 | 10 | | | | | | |
| # 8522 XYZ Dorm | | 150 | | | | | | |
| # 3701 Bldg. #8 | | | 160 | 160 | | | | |
| # 2232 Bldg. #10 - Room 201 | 20 | 20 | | | | | | |
| # 6757 Bldg. #11 - Room 20 | | 40 | | | | | | |
| # 8700 DEF hall | | | 100 | 50 | 50 | | | |
| New addition - XYZ Hall | | | | 80 | | 120 | | |
| Dean Doke's office | | | 180 | | | | | |
| Alumni Bldg. roof | | | | | | 180 | 180 | |
| Admin. Bldg. renovation | | | 30 | 160 | 160 | | | |
| Stairway West Parking | | | | | 160 | | | |
| Bldg. # 8- Rms. 101, 102, 103 | | | | | | 40 | 160 | |
| Bldg. # 12 | | | | | | | | 240 |

**Figure 3.7** Long-term schedule. (From Rex O. Dillow, ed., *Facilities Management: A Manual for Plant Administration,* 2d ed., Association of Physical Plant Administrators of Universities and Colleges, Alexandria, Va., p. 1011.)

TYPICAL WEEKLY SCHEDULE

DATE ____________ Gross Workhours Available ____________

SHOP ____________ Minus Leave Time ____________

Net Work Hours Available ____________

| Job Number | Location | Daily Workhours Scheduled | | | | | Total Hours Scheduled | Percent Completed | Comments |
|---|---|---|---|---|---|---|---|---|---|
| | | Mon | Tues | Wed | Thu | Fri | | | |
| | | | | | | | | | |
| | | | | | | | | | |
| | | | | | | | | | |
| | | | | | | | | | |
| | | | | | | | | | |
| | | | | | | | | | |
| | | | | | | | | | |
| | | | | | | | | | |
| | | | | | | | | | |
| | | | | | | | | | |
| | | | | | | | | | |
| | | | | | | | | | |
| | | | | | | | | | |
| | | | | | | | | | |
| | | | | | | | | | |
| | | | | | | | | | |

**Figure 3.8** Typical weekly schedule. (From Rex O. Dillow, ed., *Facilities Management: A Manual for Plant Administration,* 2d ed., Association of Physical Plant Administrators of Universities and Colleges, Alexandria, Va., p. 1011.)

- Weekly schedules issued for each shop reflect known absenteeism in the shop capacity figure for a given week.
- The weekly schedule shows simply the number of work hours that should be spent by each shop on each job for every day of the work week. Figure 3.8 is an example of a typical weekly schedule.
- Jobs are not scheduled until major items of materials and parts have been checked for availability.
- Weekly meetings are held for all cognizant personnel to communicate actual versus scheduled performance, and to discuss and resolve both actual and potential problems.

**Daily scheduling.** The daily scheduling interval is developed from the weekly schedule. It results when there are changes that interrupt the weekly schedule such as excessive absenteeism, materials and parts that have not arrived, emergencies, and last-minute special customer requirements. Daily schedules are normally developed by the immediate cognizant supervisor.

### Execution of work

This is the actual conduct of maintenance and repair work in response to a work request. It can be accomplished in one of two ways: in-house work force or contract. It is the point at which all the planning, estimating, and scheduling efforts pay off. Many activities (such as ordering materials and equipment; requesting and supplying support services, for example, scaffolding, land surveys, utility marking and identification; safety inspections; and developing trade and individual work schedules) are synchronized in order to achieve the common purpose of completing the work within the estimated cost and time frame.

The in-house work force should record labor hours, and material, supplies, parts, and equipment usage on a daily employee activity sheet (see Fig. 3.9). This document provides a record of worker activities for that day. It should be completed, daily, by the worker and turned in to the supervisor, who then checks the time associated with each work order. Supplies and materials usage can be annotated to this form or included on the work request.

For contracted (outsourced) work an inspector should be assigned to periodically check and inspect as the work progresses. Nothing should be placed, or approved for payment, unless it has been inspected by the department's quality control inspector.

Emergency work should be handled on a case-by-case basis. This work requires special attention. Emergencies called into the work receptionist should be verified by requesting someone (either a supervisor or craftsperson) to verify the emergency. Once verified, requests to shops, or zones, should be made via radio or pager for quick response. Emergency work takes priority over all other work.

Employee Activity Sheet

| Name: | | | Date: | | | ID #: |
|---|---|---|---|---|---|---|
| S.O. Number | Bldg | Location | Time spent | Complete? | Tag# | Customer Requestor/Contact |
| Problem: | | | | | | |
| Description/Parts/Notes | | | | | | |
| S.O. Number | Bldg | Location | Time spent | Complete? | Tag# | Customer Requestor/Contact |
| Problem: | | | | | | |
| Description/Parts/Notes | | | | | | |
| S.O. Number | Bldg | Location | Time spent | Complete? | Tag# | Customer Requestor/Contact |
| Problem: | | | | | | |
| Description/Parts/Notes | | | | | | |
| S.O. Number | Bldg | Location | Time spent | Complete? | Tag# | Customer Requestor/Contact |
| Problem: | | | | | | |
| Description/Parts/Notes | | | | | | |

Page ___ of ___ Signature: ___________

**Figure 3.9** Employee activity sheet.

## Checked and inspected

Good facility managers check! check! check! Even with continuous quality improvement programs, wherein quality is built into everything that is done, there still exists a need for inspection. Everyone in a facility department, whether they believe it or not, is part of the inspection program. Inspection is a continuous undertaking. It never ceases! Crafts persons should be trained in their trade and pride instilled in them to produce the best quality work they can. They should not be satisfied with mediocre results. Supervisors and managers should be trained the same way—to strive to continuously improve upon everything they do.

Inspections are also an important component of the PM program. It is important to note here that as PM inspections are properly accomplished; there will be a decrease in reactive maintenance and a corresponding increase in scheduled maintenance. A mechanic conducts PM of a specific piece of equipment and annotates any unusual information on the PM work ticket so that it can be entered into the CMMS PM module database. In most cases the PM inspector will verify the information in the field. If severe enough, repair can be planned and scheduled without the need for an emergency outage. As equipment ages and deteriorates, the importance of PM inspections is obvious...serious problems resulting from breakdowns can be reduced. Through planning and scheduling, downtime is reduced.

#### Closed and cost accounted

Work is properly closed out when excess materials, parts, and supplies are returned to materials control; punch list items are completed; required tests, demonstrations, or verifications have been completed; operation and maintenance manuals have been collected, reviewed, and accepted; "as built" drawings have been submitted; lessons learned are documented; labor, material, and equipment used have been assembled and added to the work order; and if reimbursable work, the customer has been formally charged, through the accounts payable system, for the work.

#### Results recorded, analyzed, and measured

Once a work order has been completed and closed out, it has to be analyzed and measured. This procedure is performed to compare actual performance against predetermined standards. As a minimum, quantities of materials used, as well as the charges that may have been credited, should be spot-checked. Actual labor hour charges should also be compared to those estimated by the planner-estimator in order to improve the value and validity of the planner-estimator's database. All of the above actions are critical in the continuous quality improvement loop. This step is important to determine if trends, lessons learned, or potential improvements can be synthesized from the work that took place. Tools that can be used to assess the effectiveness of the maintenance and PM programs can be incorporated into weekly and monthly management reports.

#### Weekly reports

- *PM inspection report.* This report is the result of follow-up inspections of work identified from PM work orders.
- *Open work order report.* This is a weekly status report of all open work orders. See Fig. 3.10.
- *Backlog report.* This is a weekly list of backlogged craft labor hours. See Fig. 3.11.

#### Monthly reports

- *Performance analysis report.* Measures the quality and quantity of work performed. Actual labor hours and materials used are compared to the estimated amount and variances are identified. See Fig. 3.12.
- *Closed work order report.* Summary of work orders closed out. See Fig. 3.13.
- *Maintenance budget report.* Includes expenses for year-to-date compared to budgeted amount and depicts variances.
- *PM maintenance summary.* This is a 12-month trend report of expenditures by equipment and system. See Fig. 3.14.
- *Trend analysis report.* This report is used to project future maintenance trends by using historical data. Forecasts are generated by graphically displaying the historical information and analyzing the data to establish trends. Typical applications include projecting future costs for materials, utilities, hazardous waste disposal, labor, and equipment; budget and financial planning; and operational planning backlog.

### Feedback

In order for a facility department to improve its service and the way it conducts business it has to receive timely and accurate feedback. This feedback, in the form of review and analy-

**Work Orders by Shop**

**Georgetown University**
**Washington, DC**

**Date range from 3/31/97 to 4/4/97**

**Page 1 of 16**
**4/10/97**

**Shop Codes:** 10
**Shop Description:** UTILITY PLANT

| Work Order # | Requestor | Building | Room | Work Order Description | Employee Assigned | Start Date | Completion Date | Total WO Cost |
|---|---|---|---|---|---|---|---|---|
| 16100 | BROWN,MONICA | INTERCULTURAL CENTER ICC_L03 | 306 | SEE NOTES | | 4/3/97 | | $0.00 |

**Work Orders printed: 1**

**Shop Codes:** 24
**Shop Description:** ZONE I MAINTENANCE

| Work Order # | Requestor | Building | Room | Work Order Description | Employee Assigned | Start Date | Completion Date | Total WO Cost |
|---|---|---|---|---|---|---|---|---|
| 15669 | DIVEN ALICE | NEW RESEARCH BUILDING NRB_Z09 | EP07 LABORATORY | AIR FLOW NOISY | BINSON J38 | 3/31/97 | | $0.00 |
| 15680 | JOHNSON LINDA | MED-DENT MDL_Z01 | 4 SE402 | INSTALL VACUUM LINES/SEE ATTAC | UCKER R98 | 3/31/97 | 4/2/97 | $14.62 |
| 15984 | DAVILA DR | MED-DENT MDL_Z01 | 4 NE420 LABORATORY | TUBE UNDER SINK BROKE/WATER F | SEAL M99 | 4/2/97 | | $0.00 |
| 16016 | APPERSON, CHERYL | NEW RESEARCH BUILDING NRB_Z09 | G00C1 CORRIDOR | ADJUST SWINGING DOORS/SEE NOT | | 4/2/97 | | $0.00 |
| 16024 | KOZIKOWSKI ALAN | NEW RESEARCH BUILDING NRB_Z09 | EP07 LABORATORY | HANG MARKER BOARD IN OFFICE | | 4/2/97 | | $0.00 |
| 16077 | SINGLETON, MARY | LEAVEY CENTER LEA_P02 | S-316 RECEPTION | DR BELL NOT WORKING | PRATT D17 | 4/3/97 | | $0.00 |
| 16096 | JENNINGS JASON | LEAVEY CENTER LEA_P02 | 1319 VITAL VITTLES | REMOVE FREON FROM COOLER/SEE | ILLIAMS R53 | 4/3/97 | | $0.00 |
| 16153 | HENDRICKS CECILA | LEAVEY CENTER LEA_P02 | S-530 MAIL ROOM | HANG PLAQUE ON WALL | PRATT D17 | 4/3/97 | | $0.00 |
| 16174 | MICHELLE | NEW RESEARCH BUILDING NRB_Z09 | E202A OFFICE | TOO HOT | COATS T58 | 4/4/97 | | $0.00 |
| 16199 | MUNOZ, JOE M | NEW RESEARCH BUILDING NRB_Z09 | B0M1 MECHANICAL ROOM | REPAIR BOOSTER PUMP | COATS T58 | 4/4/97 | | $0.00 |

**Figure 3.10** Weekly status report, open work orders.

**WO Backlog Report by Building**

**Page 1 of 6**

**4/10/97**

**Georgetown University**

**Washington, DC**

***Date range from 11/1/96 to 4/10/97***

**Building Code:** COP_K01

**Building** COPLEY

| Days Late | Work Order # | Date WO Issued | Date WO Required | Room Code Description | Building System | Requestor/ Job Description | Employee Assigned | Vendor Assigned |
|---|---|---|---|---|---|---|---|---|
| | 12168 | 2/25/97 | | 516<br>LOUNGE | MECH | REPLACE FCU CONTROL SWITCH | | |
| | 12161 | 2/25/97 | | | CARP | install window stops | | |
| | 15893 | 4/2/97 | | B-25<br>BOILER ROOM | 15730 | | | |
| | 15614 | 3/31/97 | | OUT<br>OUTSIDE | 16210 | | | |
| | 15432 | 3/31/97 | | B-24<br>LAUNDRY | | | | |
| | 15590 | 3/31/97 | | B-17 | 15510 | | | |
| | 13459 | 3/10/97 | | B-25<br>BOILER ROOM | 15250 | | | |
| | 14776 | 3/27/97 | | | CARP | REPLACE MISSING & CRACKED LENSE | | |
| | 13854 | 3/17/97 | | 530<br>BEDROOM | MECH | broken fcu speed switch | | |
| | 15604 | 3/31/97 | | B-17A<br>MECHANICAL ROOM | 15510 | | | |
| | 9657 | 2/4/97 | | B-25<br>BOILER ROOM | 15730 | | | |
| | 15520 | 3/31/97 | | B-17 | 15510 | | | |

**Figure 3.11** Backlog report.

| Week Ending | Work Order # | Estimated Hours | Estimated Material Cost | Actual Hours | Actual Material Cost | Labor Estimating Efficiency (1) |
|---|---|---|---|---|---|---|
| 13-Mar | 72332 | 6.2 | 140.23 | 9.4 | 120.98 | 51.60% |
| | | | | | | |
| | | | | | | |
| | | | | | | |
| | | | | | | |
| | | | | | | |
| | | | | | | |
| | | | | | | |
| | | | | | | |
| | | | | | | |
| | | | | | | |
| | | | | | | |
| | | | | | | |
| | | | | | | |
| | | | | | | |

Performance Analysis Report

1) (Estimated Hours - Actual Hours) / Estimated hours X 100 = %

**Figure 3.12** Performance analysis report.

sis, provides the information required to ensure that actions are being performed as planned. It also allows the facility manager to focus efforts for organizational improvement. For instance, the facility manager, and his or her staff, should meet periodically (quarterly may be opportune) to evaluate the efficiency of conducted operations, review the quality of performance, measure the performance of management initiatives, and identify deficiencies and opportunities for improvement. Major accomplishments should be highlighted and disclosed to higher management. Remember: The "boss" doesn't necessarily know what you, the facility manager, are doing unless you tell him or her; or unless a critical organization operating problem has been forwarded by a totally unhappy organization department head. The review and analysis process should be formalized and documented to ensure its success and continued emphasis. The facility manager has to form a viable department planning group that will develop strategy and provide the necessary leadership to the rest of the department. The results of this program should

*Work Order Status Report*

| Georgetown University<br>Washington, DC | *Date range from* *4/1/97* *to* *4/10/97* | Page 1 of 14<br>4/10/97 |
|---|---|---|

| Work Order # | Current Date | Work Order Description | Status |
|---|---|---|---|
| 15766 | 4/1/97 | LIGHT OUT 3RD FL HALL NEAR MENSRM | Complete |
| 15845 | 4/1/97 | RAIL DOWN IN 91-96 | Complete |
| 15805 | 4/1/97 | HANDLE BROKEN/KEY WONT TURN | Complete |
| 15755 | 4/1/97 | MIRROR BROKEN ON CABINET DR | Complete |
| 15820 | 4/1/97 | LATCH/BAR/ SCREWS OUT OF DOOR | Complete |
| 15812 | 4/1/97 | ORDER TOOLS | Complete |
| 15741 | 4/1/97 | REPLACE DOOR STOP-1ST FL MEN'S ROOM | Complete |
| 15850 | 4/1/97 | LECTURN HAS NO LIP | Complete |
| 15753 | 4/1/97 | CLEAN LAUNDRY ROOM FILTER | Complete |
| 15830 | 4/1/97 | SEE NOTES | Complete |
| 15777 | 4/1/97 | PLEASE CHECK DISHWASHER SEE IF WE | Complete |
| 15765 | 4/1/97 | 10,LIGHTS OUT ON 3RD FLOOR | Complete |
| 15780 | 4/1/97 | REPAIR DOOR CLOSER | Complete |
| 15762 | 4/1/97 | POLE LIGHT OUT | Complete |
| 15746 | 4/1/97 | DRAIN AIR COMPRESSOR | Complete |
| 15852 | 4/1/97 | LIGHT OUT IN BATHROOM | Complete |
| 15782 | 4/1/97 | COIL ON SERVING LINE REACH IN FROZE | Complete |
| 15758 | 4/1/97 | OUTLETS NOT WORKING | Complete |
| 15781 | 4/1/97 | REPAIR DAMAGES DUE TO BREAK IN | Complete |
| 15742 | 4/1/97 | TOILET CLOGGED | Complete |
| 15779 | 4/1/97 | PLUMBER TO DISCONNECT/SEE BUZZ | Complete |
| 15840 | 4/1/97 | GARBAGE DISPOSAL CLOGGED | Complete |
| 15817 | 4/1/97 | TOILET CLOG | Complete |
| 15822 | 4/1/97 | COMPRESSOR LOCKED UP/W.COOLER/LADY | Complete |
| 15745 | 4/1/97 | DRAIN AIR COMPRESSOR | Complete |
| 15833 | 4/1/97 | 5TH FL LADIES RM FAUCET BROKEN | Complete |
| 15839 | 4/1/97 | BRICK MISSING FROM OUTSIDE WALL | Complete |
| 15841 | 4/1/97 | COLD | Complete |
| 15769 | 4/1/97 | CHANGE LOCK AND CUT 6 KEYS | Complete |
| 15844 | 4/1/97 | TOILET CLOGGED | Complete |
| 15752 | 4/1/97 | LIGHTS OUT | Complete |
| 15771 | 4/1/97 | BATHRM LIGHT OUT | Complete |
| 15835 | 4/1/97 | CK COURTYARD DR/NOTLOCKING BEHIND | Complete |
| 15829 | 4/1/97 | CUT (2) KEYS #45 FOR SUPPLY CLOSET | Complete |
| 15757 | 4/1/97 | REPAIR HOLES ON WND FLOOR HALLWAY | Complete |
| 15848 | 4/1/97 | CALLED LOCKSMITH IN 4:00A.M. | Complete |
| 15759 | 4/1/97 | POWER OUT | Complete |
| 15790 | 4/1/97 | FRONT DOOR GLASS COMING OUT | Complete |
| 15842 | 4/1/97 | BATHROOM LIGHT OUT | Complete |
| 15813 | 4/1/97 | REMOVE WATER COOLERS/STORE IN VIC | Complete |
| 15748 | 4/1/97 | DRKNOB OFF DOOR CANT GET INTO RM | Complete |
| 15749 | 4/1/97 | CLOGGED SHOWERS-GROUND, 1ST,2ND FL | Complete |
| 15828 | 4/1/97 | LIGHT OUT | Complete |
| 15778 | 4/1/97 | 7TH FL STORAGE CLOSET LOCK BROKEN | Complete |

**Figure 3.13** Closed work order report.

**Work Orders by Shop**

**Georgetown University**
**Washington, DC**

*Date range from* **4/1/97** *to* **4/10/97**

**Page 1 of 5**
**4/10/97**

**Shop Codes:** 20
**Shop Description:** MAINTENANCE CONTROL

| Work Order # | Requestor | Building | Room | Work Order Description | Employee Assigned | Start Date | Completion Date | Total WO Cost |
|---|---|---|---|---|---|---|---|---|
| 15855 | | LXR<br>LXR_A01 | G00U3<br>MECHANICAL ROOM | BI-MONTHLY LESLIE READINGS | ANTER H55 | 4/2/97 | 4/2/97 | $0.00 |
| 15856 | | LXR<br>LXR_A01 | G00U3<br>MECHANICAL ROOM | BI-MONTHLY LESLIE READINGS | ANTER H55 | 4/2/97 | 4/2/97 | $0.00 |
| 15857 | | COPLEY<br>COP_K01 | B-25<br>BOILER ROOM | BI-MONTHLY LESLIE READINGS | ANTER H55 | 4/2/97 | 4/2/97 | $0.00 |
| 15858 | | COPLEY<br>COP_K01 | B-25<br>BOILER ROOM | BI-MONTHLY LESLIE READINGS | ANTER H55 | 4/2/97 | 4/2/97 | $0.00 |
| 15859 | | POULTON HALL<br>PLH_D19 | B07<br>BOILER ROOM | FILTER INSPECTION FOR CLOSED LO | | 4/2/97 | | $0.00 |
| 15860 | | LXR<br>LXR_A01 | G00U3<br>MECHANICAL ROOM | FILTER INSPECTION FOR CLOSED LO | | 4/2/97 | | $0.00 |
| 15861 | | LXR<br>LXR_A01 | G00U3<br>MECHANICAL ROOM | FILTER INSPECTION FOR CLOSED LO | | 4/2/97 | | $0.00 |
| 15862 | | WALSH<br>WAL_A04 | 100M<br>MECHANICAL ROOM | FILTER INSPECTION FOR CLOSED LO | | 4/2/97 | | $0.00 |
| 15863 | | NEVILS<br>NEV_A05 | B-02B<br>MECHANICAL ROOM | FILTER INSPECTION FOR CLOSED LO | | 4/2/97 | | $0.00 |
| 15864 | | MED-DENT<br>MDL_Z01 | BSE B03A<br>MECHANICAL ROOM | FILTER INSPECTION FOR CLOSED LO | | 4/2/97 | | $0.00 |
| 15865 | | LAUINGER LIBRARY<br>LLI_E01 | B-15E1<br>MECHANICAL ROOM | FILTER INSPECTION FOR CLOSED LO | | 4/2/97 | | $0.00 |
| 15866 | | LAUINGER LIBRARY<br>LLI_E01 | B-15E1<br>MECHANICAL ROOM | FILTER INSPECTION FOR CLOSED LO | | 4/2/97 | | $0.00 |
| 15867 | | PRECLINICAL SCIENCE<br>PCS_Z05 | B09<br>EQUIPMENT ROOM | FILTER INSPECTION FOR CLOSED LO | | 4/2/97 | | $0.00 |
| 15868 | | PRECLINICAL SCIENCE<br>PCS_Z05 | B09<br>EQUIPMENT ROOM | FILTER INSPECTION FOR CLOSED LO | | 4/2/97 | | $0.00 |

**Figure 3.14** PM maintenance summary.

provide information and data for each manager and supervisor's annual performance evaluation.

Following is a review and analysis suggested agenda:

- Update on open action items from the previous meeting (review of minutes from previous meeting).
- Discuss available resources versus required and make adjustments.
- Review goals and objectives and the obstacles to meeting them. Alternatives for corrective action are discussed.
- Assign responsibilities and suspense dates for solving identified problems.
- Review quality improvements and accomplishments.
- Outline topics and objectives for the next meeting.

The following should be included in a review and analysis meeting:

- Presented information and discussion should be captured in minutes of the meeting.
- Minutes should include date, time, attendees, discussion items, highlights of policy changes, customer feedback information (percentages and comments), open items, and managers and supervisor's accomplishments.
- Successful quality improvement actions, and other accomplishments, should be mentioned.

#### Customer feedback

Customer feedback can be solicited in various ways. Mechanics can leave "while you were out" cards when they respond to a service order and the initiator is not available. See Fig. 3.15. By leaving this card, the initiator knows that the facility department responded and either repaired the deficiency or defined a reason why it was not fixed. The key here is that customers are satisfied if they know what's happening. They only get disgruntled when they're "kept in the

**Georgetown University**
**Facilities Management**

**While you were out,**
**our staff had to enter your area to:**

______________________________

Date: ________ Time: ________ Zone # ________

Employee Name: ________ Shop # ________

Work Completed: Yes/No Work Order # ________

Comments: ________

Inquiries: (687-3432)

**Figure 3.15** While you were out card. (These cards can be the size of 3 × 5 cards, which allows them to be carried in a shirt pocket, or be made in the form of 3 × 5 Post-it® notes so they can be adhered to a refrigerator, door, desk, etc.)

dark." A second means of feedback is for the work reception center to call back a sampling of customers daily, following up on completed work orders and soliciting feedback. In this case it would be worthwhile to standardize questions. These questions will provide immediate feedback and serve as a good measure of customer perceptions of the services they receive. These questions can be as simple as:

- Was the work completed to your satisfaction?
- Were facility department personnel courteous?
- Was a "while you were out" card left?

Another, more formal, means of obtaining customer feedback is to use a formal survey. This technique will be covered in detail later in this chapter.

## Building Operational Plan

Facility managers are responsible for overseeing the safe, efficient operation and maintenance of each facility under their control. As such, their decisions impact every member of that organization every day. Within the numerous processes they are answerable for there are common issues: how to schedule and coordinate work efficiently, how to control startups and shutdowns of equipment and operating systems, how to handle emergency situations, how to coordinate diagnosis and repair of trouble, and how to use benchmarking and performance indicators to determine the operational status of equipment and systems.

### Shutdowns and start-ups

**Shutdowns.** Scheduled shutdowns are not considered emergencies. Advance notification to customers should be provided in writing at least 72 hours prior to the scheduled outage. Alternative support may be required if the shutdown will last more than several hours. This support could consist of lighting, heat, potable water, or cooling and can be planned for, and connected, in advance. Typical planned shutdowns include emergency generator testing; utility repairs; equipment repair, replacement or overhaul; sprinkler and fire protection system testing and repair; and building heating, ventilating, and air-conditioning (HVAC) system repair or adjustment.

**Start-ups.** The objective is to get equipment operating in an efficient, safe, cost-effective manner as quickly as possible. The following points should be considered:

- Expect equipment problems. Depending on how long the equipment has been down, fluids and lubricants will have drained from metal surfaces, systems will become air bound, and equipment will lose calibration.
- Be cautious of pressure systems. Because of no demand, systems that are normally pressurized may have higher than normal pressures.
- Ensure that all necessary materials and supplies are on hand. This will reduce workforce idle time and provide for a more balanced workload.
- Ensure that prior to start-up all maintenance and utility crews are on hand early, and available to minimize and respond to problems.

### Emergency situations

Emergency situations require immediate action. When compared with scheduled maintenance, emergency maintenance is a very small percentage (usually less than 10 percent). The detailed planning and scheduling that precede scheduled maintenance are impractical in emergency situations because of the immediacy of the situation. This is why emergency situations must be planned ahead of time and written procedures put in place. This will be covered in more detail following. As a minimum, the facility manager should plan for three areas of immediate concern.

### Fire protection

- The first step is to develop a fire protection protocol, unique for each building. It should define who the "authority having jurisdiction" is (prior to the fire department arrival): this could be the building security officer, the property manager, or the facility manager; building evacuation procedures; description of the alarm system and location of annunciator panels, sprinkler valves, and fire pump; shutdown procedures; emergency generator location and description; and what to do if the systems fail.
- Training is extremely important. Occupants of buildings should be familiar, through periodic fire drills, with how to evacuate and also know location and how to activate alarm pull stations. Maintenance and operations personnel should be trained, semiannually, on the location of specific monitoring panels and shutoff valves and disconnects and should have the protocol etched in their minds.
- Designate a spokesperson to coordinate public relations and news media issues and statements.

### Power outage

- Control devices should always be set for "fail-safe."
- All electrical start-stop switches should be shut off. This will prevent damage to certain types of electrical equip-

ment should partial electrical power be restored. Also, circuit breakers may switch off if they sense an overload when power is restored.

- Verify that all emergency systems are operational. If the outage lasts for several hours, check the fuel tanks of emergency generators.
- If in the winter season, check throughout the building for possible pipe freeze-ups. Crack open water faucets at various locations throughout the building.

### Flooding

- If the building is located in a flood plain, contact the state flood control agency or the U.S. Army Corps of Engineers to obtain information on expected time of arrival and water levels.
- Provide property protection. Time permitting, all doors and windows should be sealed and sandbagged. Emergency generators should be obtained to operate pumps and provide emergency lighting. Electrical equipment should be de-energized. Where feasible, delicate equipment should be relocated to safe locations.
- Implement procedures that define who makes decisions to evaluate and shut down equipment, details evacuation procedures, and determines emergency crews.

### Trouble diagnosis and coordination

As mentioned before, coordination is extremely important. You can have the most competent mechanics and the best maintenance plan, but if there is no coordination, the facility department is looked on as incompetent.

**Diagnosing a maintenance problem.** Diagnosis is the identification of maintenance problems causes and effects. This initial step eventually leads to correction and elimination of such causes as misapplication and operator abuse. The "diagnostician" plays an important role here. This individual must have the technical background and personality to be able to find a problem and teach mechanics and operators the proper maintenance techniques without coming across as a "know-it all" or trying to find fault. It is not easy to find an individual with these characteristics. Before any diagnosis of a maintenance problem can begin, the technician must have the proper training and experience, understand the steps involved in diagnosing a problem, and be knowledgeable and skilled in the use of diagnostic methods and instruments. In some cases it may be more cost-effective to contract for this service.

### Coordination with third parties

If customers have been impacted by an equipment outage or if they are going to pay for the repair, they must be told what happened, why it happened, and the corrective action taken to prevent it from happening again. Unless this is done, the facility department loses credibility. Even the best operating facility department, having an excellent PM program to minimize equipment downtime and maximize scheduled maintenance, will experience occasional emergencies wherein customers cannot be warned of unscheduled outages. This is where good customer relations comes into play.

### Benchmarking

The purpose of benchmarking is to improve an organization's performance by comparing its practices to those of another. In order for facility managers to conduct a bench marking study, it is important that they know their own processes. This is necessary to get the maximum value from study comparisons with other facility organizations. Good preparation is important before a study is begun! Benchmarking equipment and systems in individual buildings are important in order to measure performance and make comparisons of that equipment and system with other organization equipment and systems.

### Statistical process control

Once thorough understanding of a procedure is attained, major advances in service quality, productivity, and cost can be achieved. To develop this knowledge a tool such as statistical process control (SPC) can be used. SPC is used in industrial and manufacturing environments as a process to improve quality and minimize waste. It is a statistical technique suited for use whenever something can be measured or counted in order to show trends that will guide future actions. It is typically used in continuous improvement processes and can be creatively adapted to various facility building operating systems. Facility managers can use SPC to help better understand their processes, measure and analyze quality, and ultimately improve maintenance.

### Comprehensive facility operational plans

All facility managers have the responsibility to effectively plan for and respond to an emergency situation and bring that situation to a suitable conclusion. To do this preparation is vital! The establishment of a formal written plan which defines how to handle emergency procedures is essential. This plan should be disseminated to everyone who has responsibility to put emergency procedures into practice. Additionally, emergency drills and problem exercises should be periodically scheduled to ensure that facilities personnel are properly trained and conditioned to react quickly and skillfully when an emergency situation occurs.

### Emergency response plan

Emergencies occur! A good emergency response plan defines duties and responsibilities, identifies various emergencies which may occur, explains what to do before, during, and after an emergency, and details the procedures for resolving most conceivable emergencies.

**Organization.** An emergency coordinator should be appointed by top management. This individual should be assigned to a central location [emergency operations center (EOC)]. He or she should coordinate actions during both the emergency and the restoration period. The EOC is the primary control point for the coordination and handling of responses to emergencies. Staffing of the EOC is dependent on the type, magnitude, and location of the emergency.

**Concept of operation.** The concept of operation is a statement of the emergency coordinator's visualization of how an emergency should be handled from start to finish. It should be stated in sufficient detail to ensure appropriate action. After normal work hours, the organization's security office should be the first office to be notified of an emergency. This is due to several reasons: security officers patrol buildings and can physically detect a problem, alarm systems (fire and environmental) terminate at the security office, or individuals detecting an emergency situation call the security office to initiate the emergency response plan.

**Command and control.** During major emergencies, a positive chain of command must exist and be functional. There must be a clearly identified chain of organization as shown following:

- *Command.* The responsibility for resolving any emergency situation rests with top management. However, the emergency coordinator should be the executor of the emergency response plan, and should assist top management in planning, training, responding to, and mitigating emergencies.
- *Responsibility succession.* In the absence of the final top management responsible individual there exists a need to define a successive line of authority and responsibility for decision making.
- *Control.* Initial control responsibility for emergency situations rests with the security manager. As a minimum, this responsibility should include the following activities:

  Immediate evacuation of affected buildings necessary for the life safety of occupants.

Initial response to emergencies in the facility or building. This response determines the type and size of the emergency.
Liaison coordination with external support agencies such as police and fire departments.
Exterior crowd control at all emergency scenes.
Assisting the emergency coordinator with initial activation of the EOC.

**Communications.** Adequate, effective, and redundant communications systems must be available to effectively respond and control emergency situations. These systems must be an integral part of the EOC. They must be reliable and capable of functioning during periods of power loss.

**Types of emergencies.** Depending on the type of facility, or property, the following listed emergency situations should be addressed. For each of these emergencies the plan should address preparations before an emergency occurs, training requirements, response procedures (which include description of personnel actions for each credible emergency requiring specialized response), and restoration procedures.

- Hazardous material emergencies
- Fire emergencies
- Natural disaster emergencies
- Bomb threat emergencies
- Utility outage emergencies
- Labor unrest emergencies

**Support services.** These are services that can be provided to measure, control, or mitigate emergency situations. Information that should be incorporated here includes:

- Resources available
- Temporary housing support
- Updated list of telephone numbers and points of contact
- Local building codes and regulatory requirements

- Insurance information
- Flowchart of the chain of command
- Reporting procedures and required documentation
- Federal, state, and city disaster assistance procedures
- Information on how to track costs

**Guidelines for facility specific plans.** Specific emergency procedures for each individual facility should be prepared. The plans should contain, as a minimum, the following information:

- Brief description of the facility or type of operation conducted
- Procedures for evacuating and accounting for any visitors and for all personnel normally working in the facility to include:

  Determination of when evacuation is necessary
  Designation and use of assembly areas to account for personnel
  Selection of evacuation routes out of the facility
  Establishment of a personnel accountability system
  Procedures to evacuate physically impaired personnel
- Description of the means of communicating the emergency situation to all personnel
- Identification of specialized equipment needed by units responding to the emergency
- Procedures for shutting down all utilities to the affected area and securing the facility from unauthorized entry
- Floor plans and blueprints
- Building equipment and systems information

### Hazardous materials plan

Hazardous and toxic materials are all materials (gaseous, liquid, or solid) that can cause physical or chemical changes in the environment, affect property, or affect the health or physical well-being of living organisms. A release or spill of hazardous material requires the controlling organization to immediately notify people in the surrounding vicinity that a release or spill has occurred. If deemed necessary, an evacuation of the area (room, floor, or building) will be accomplished. A specific hazardous material concern to all facility managers is the use of refrigerants for cooling and refrigeration equipment. In accordance with international agreements (Montreal Protocol), and federal laws, chloride-bearing refrigerants (CFCs) were phased out of use on January 1, 1996. Phase-out of HCFCs will begin in 2003.

The purpose of a hazardous materials plan, then, is to:

- Define types of releases or spills.
- Identify procedures for personnel to follow at the scene.
- Define cleanup procedures (including contamination cleanup).
- Describe and identify spill management equipment and materials needed.
- Describe transportation equipment requirements.

**Training.** Facility managers are not only responsible for the facilities they maintain, but they also have an obligation to ensure their personnel are appropriately trained in hazardous materials usage in the workplace. The following training exercises should be considered:

- Hazards communication (right-to-know)
- First responder awareness
- First responder operations

**Assessment.** According to the Code of Federal Regulations, 29 CFR 1910.132, employers are required to perform a workplace condition assessment to determine whether hazards are present, or likely to be present, that necessitate the use of personal protective equipment (PPE). If such hazards exist or are likely to, the employer must select appropriate

PPE and certify that a workplace hazard condition assessment was performed.

### Refrigerant management

**Refrigerant accountability.** According to the Environmental Protection Agency (EPA) Stratospheric Ozone Protection, Final Rule Summary, EPA-430-F-93-010, "Technicians servicing appliances that contain 50 or more pounds of refrigerant must provide the owner with an invoice that indicates the amount of refrigerant added to the appliance. Technicians must also keep a copy of their proof of certification at their place of business. Owners of appliances that contain 50 or more pounds of refrigerant must keep servicing records documenting the date and type of service, as well as the quantity of refrigerant added." Essentially, refrigerant must be accounted for from the day of purchase to the day of disposal.

Many computer software packages are available on the market designed to provide a means of tracking refrigerant usage in accordance with EPA requirements. These software packages provide the capability of matching usage in relation to an asset's basic refrigerant charge, which can help in determining leakage. An example of a refrigerant usage log can be seen in Fig. 3.16. To protect themselves and their organization, facility managers should establish a policy for purchasing, accounting for, and reclaiming refrigerant, and then develop a system to implement that policy. This is essential in the event the EPA decides to conduct an audit.

**Technician certification.** As of November 14, 1994, federal law requires that all mechanics working on air-conditioning or refrigeration equipment be certified to correctly service that equipment, including recovery and recycling practices. These mechanics are required to pass an EPA-approved test. There are four types of certification:

- Type I: Servicing small appliances.
- Type II: Servicing or disposing of high-pressure appliances.
- Type III: Servicing or disposing of low-pressure appliances.

**GEORGETOWN UNIVERSITY REFRIGERANT USAGE LOG**

Technician Name:__________________ Cylinder Number:____________

Issue Date:_________ Closed Date:____________

Refrigerant Type:________ Starting Weight:____________

| Date | SPAN Ticket # | Equipment ID | Refrigerant Cylinder # | Refrigerant | | | Total Refrigerant | |
|---|---|---|---|---|---|---|---|---|
| | | | | New | Recovered | Contaminated | Added | Removed |
| | | | | | | | | |
| | | | | | | | | |
| | | | | | | | | |
| | | | | | | | | |
| | | | | | | | | |
| | | | | | | | | |
| | | | | | | | | |
| | | | | | | | | |
| | | | | | | | | |
| | | | | | | | | |
| | | | | | | | | |

Total New Refrigerant Used:_________ Cylinder #________
Total Recovered Refrigerant:_________ Cylinder #________
Total Recovered Refrigerant:_________ Cylinder #________
Total Recovered Refrigerant:_________ Cylinder #________
Amount Recovered from new Cylinder :__________
When Refrigerant was added was the reason Unintentional Venting?Yes / No (circle one)
Did you repair all leaks according to EPA 608? Yes /No (circle one)

**Please Weigh all Cylinders before and after each use. Keep all Cylinders locked! Accurate reports are very important, please make sure that all new cylinders are emptied and add up to starting weight.**

Technician Signature:____________________Date:______________

**Figure 3.16** Refrigerant usage log.

- Type IV: Servicing all types of equipment (universal).

Under Section 609 of the Clean Air Act, sales of CFC-12 refrigerant in containers smaller than 20 pounds are now restricted to technicians certified under EPA's Motor Vehicle Air Conditioning regulations.

### Safety plans

All organizations want to reduce the loss of downtime, lost production, and Worker's Compensation claims related to accidents. In order to improve safety within an organization and, hopefully, decrease the number of accidents and injuries on the job, a formal written safety plan is needed. The safety plan provides a system to constantly monitor the work environment to minimize potential safety and health threats.

**Purpose.** The safety plan's purpose is to establish standard procedures for normally occurring activities. These procedures, when followed, save the organization money by keeping safety continuously in the minds of everyone.

**Involvement.** In order for employees to "buy into" the safety plan they have to see the policy put into practice. This requires that safety be included in the organization's mission statement; various safety topics, germane to the functions of the organization, should be included in the plan; and safety training; consistent enforcement of safety policies and procedures; and positive reinforcement of safe working behavior should be included. Listed below are a variety of methods used by organizations to promote safe working areas and prevent accidents.

- Area involvement in the process. It is very important to have input from all areas affected by the safety topic. Office areas, the loading dock, production areas, distribution areas, mail room, supply and storage areas, laboratory areas, mechanical rooms, and others should be included.
- Safety and health committees. These committees are formed to serve several basic functions.

  Create and establish safety guidelines.
  Plan and implement facility safety programs.
  Establish and review safety training programs.
  Investigate accidents.
  Conduct and evaluate safety inspections.
  Initiate ideas and plans to correct safety problems.
  Hold weekly and monthly safety meetings.
  Develop a safety suggestion program.
  Design safety training sessions with employee input.

Stress the organization's long-term safety goals in light of employees' involvement.
Institute and communicate an organization philosophy that accidents are not acceptable and will not be tolerated.
Use statistical techniques to categorize accidents into systematic shortcomings and human error.

**Accident reporting.** To facilitate a uniform response to accident investigation and provide for the safety and health of employees a standardized accident investigation report form should be implemented. This form should be completed by the appropriate supervisor, within a specified time period after the accident occurs, for any accident or on-the-job injury. The facility manager should review each form for completeness, determine what happened, and take corrective action to prevent similar accidents in the future.

### Fire protection plans

The facility manager normally has responsibility for fire protection and prevention within a facility. This responsibility can be divided into two categories: (1) regular prevention and protection inspections, and (2) how to control fires that occur. A planned fire prevention and protection program must provide for the maximum protection against fire hazards to life and property consistent with the mission, sound engineering, and economic principles. Additionally, facility managers should have a plan to combat fire emergencies.

**Prevention and protection.** Inspections should be conducted monthly and are performed to reduce the potential of a fire hazard as shown following:

- *Prevention.* Environmental situations known to be causes of fires are inspected first. These situations include storage and handling of flammable materials and liquids; operation of electrical equipment and associated wiring; high-temperature generating equipment, such as hot work involving welding and soldering equipment; smoking noncompliance; and finally, proper housekeeping procedures.

- *Protection.* Inspection and testing of fire protection equipment are intended to ascertain the operability of that equipment. This inspection should identify and verify that each fire protection control valve functions correctly; verify the service availability of water supply; complete a fire pump checklist for each pump; examine the critical components of special extinguishing systems; test all sprinkler systems and alarms; check all fire extinguishers; identify each hydrant; determine that all fire doors work properly; verify that smoke and heat detectors operate properly; and finally, make sure that protective signaling devices function correctly.

**Fire control.** Upon discovery of a fire, the fire alarm should be sounded. If the fire is small and an extinguisher is nearby, the fire should be extinguished with the extinguisher. If the fire is larger and spreading, clear the building. If time permits,

- Close all doors and windows.
- Turn off all oxygen and gas outlets.
- Disconnect electrical equipment.
- Turn off all blowers and ventilators.

**Plan.** Fire emergencies can occur at any time. Having a fire plan developed is good insurance against total chaos. As a minimum the fire plan should cover the following:

- Authority having jurisdiction. Define the overall organization having ultimate jurisdiction until the fire department arrives.
- Define what measures should be taken before, during, and after the fire emergency.
- Training requirements.

- Formation of an organization fire brigade.
- Ensure plans and drawings for each building are available. These plans should identify electrical, chilled water, steam, telephone and data cable locations, and other utility distribution systems.
- Specific requirements should be spelled out for the operations and maintenance staff, contracts section, purchasing staff, and public relations staff.
- Requirements for fire watch.

### Labor unrest

Facilities managers having a unionized workforce will have unique challenges facing them in the event of a labor strike or threat of any job action. A strike or walkout can leave a facility critically short staffed, vulnerable, and unable to deal with a crisis. Consequently, a detailed strike contingency plan should be prepared ahead of time. This plan should include sections on:

- Staff members, vendors, suppliers, contractors, utility suppliers, and police and fire departments. They need to be kept informed of the status of the strike. Customers must also be informed of the impact to them. Using voice and electronic mail broadcasts may be the best way to keep the staff and customers informed of the following:
- Services that would:

  Continue to be provided.
  Be curtailed or suspended.
  Be contracted out.
- Staff reassignments to cover critical responsibilities and vital services.
- Protection of facility equipment and personnel.
- The organization's security department. It must be given specific instructions on how to respond to labor unrest situations.

- A detail log including date, time, and description of the incident. This should be kept in the event later legal action is necessary.

### Facility occupant support plan

Today, one of the biggest challenges facing facility managers is to provide high-quality customer service. To do this, facility managers must ensure that not only routine scheduling of work is conducted, but also that it is done in a caring manner. High-quality service exists only when the expectations of customers are exceeded.

Many facility departments believe they provide excellent service because all the mechanical and electrical systems in a building operate and meet required life safety and building codes, burned-out lights are replaced immediately, maintenance and repair backlog work has been reduced, and service technicians respond quickly when a service request is submitted. All of this may be true and still customers may not be satisfied because their expectations are not exceeded. Possible cause for this may be the way the customers perceive the service that is being provided.

### Improving perceptions

There must be caring and consistency in the way service is provided before customers will perceive facility support as efficient, effective, and customer-oriented. This change will come each time they have a positive experience as a result of the service being provided. Expectations must be exceeded every time. Every service opportunity must be seized to influence customer perceptions in a positive manner. Below are some ideas that will help.

- Minimize interference with customer activities.
- Curtail unplanned outages. Ensure customers have, as a minimum, 72 hours notice.

- Coordinate with customers, just to provide them information. They will accept almost anything as long as they know the impacts and why it has to be.
- Provide proper and timely service for customer work requests. As discussed above, work requests should be screened, evaluated, and then prioritized. Time frames should be established for each level of work. Assigned priorities must be adhered to. A system should be established to "close the loop" with customers if work cannot be accomplished in a timely manner. Electronic mail lends itself nicely to doing this.
- Do not let problems that occur go unanswered. If a facility manager learns of a problem wherein a customer is dissatisfied, investigate the reasons on both sides to "close the loop" with the customer. If the facilities department erred, acknowledge that, apologize, and let the customer know that you will implement corrective action....But respond to the customer! Again, customers will accept almost anything as long as they know the department is trying to improve and that it does care. Michael LeBoeuf, in his book *How to Win Customers and Keep Them for Life* (Putnam's Sons, 1987), states "A rapidly settled complaint can actually create more customer loyalty than would have been created if it had never occurred. Customers are much more likely to remember the 'extra touch,' fast action, and genuine concern that you exhibited when they felt dissatisfied."

### Determining wants

To exceed customer expectations facility managers must determine what customers need and what they expect as shown following:

- *Need.* One effective method of determining need is to go see the customer, face to face. Personal contact has a way of cementing relationships and developing rapport.
- *Expectations.* To determine customer expectations a formal written survey is a good technique to use. Once you

have it developed, plan on distributing it annually. Start a mailing list of persons to mail the survey to and keep adding to it. Include people who have experienced problems with service in the past; budget and financial people; and key department, floor, or building coordinators.

Remember the survey has to be simple in order for people to take the time and interest to complete it. They do not want to spend a lot of time composing and writing responses, but they will check a box corresponding to a yes/no question. It should take them *less than 5* minutes to complete. A sample survey is shown in Fig. 3.17. Returning the survey has also got to be simple. A card-type survey, preaddressed, with postage applied (if necessary) can be used once the person filling out the survey is finished, that individual can just drop it in the mail (campus mail, postal system, or building/department mail system). In addition to surveying external customers, try issuing an annual survey to the facility department workforce. This is a good way to take the pulse (measure morale) of the organization.

### Provide feedback

Most customer complaints and dissatisfaction are generated because of a lack of communication. Implementing a simple customer communications program will eliminate many routine complaints before they mushroom into major customer relations problems. Below are some ideas that can be used.

- *Call-backs.* On a daily basis have the work receptionists make call-backs to customers where work they requested has been completed. A good percentage to use would be to call back 20 percent of the work orders that have been completed. This technique lets 20 percent of the customers who called in a problem know that the facility department measures its performance, identifies follow-up problems which can then be proactively corrected, and finally provides a daily gauge to measure customer satisfaction.

**Customer Survey**

***Please indicate your level of satisfaction by circling 1, 2, 3, 4, or 5, 5 being the highest and 1 being the lowest, 3 being "satisfactory".***

**1. Are your common areas maintained in a generally clean and attractive manner by our custodial staff?**

**1 2 3 4 5**

**2. Did we respond satisfactorily to your request for service?**

**1 2 3 4 5**

**3. Was a requested repair or correction completed to your specifications?**

**1 2 3 4 5**

**4. Was the work area left in a clean and proper manner?**

**1 2 3 4 5**

**5. Was the worker involved courteous in manner and professional in appearance?**

**Figure 3.17** Customer survey.

1 2 3 4 5

**Please give us any additional comments and suggestions.**

________________________________________________

________________________________________________

________________________________________________

________________________________________________

________________________________________________

________________________________________________

________________________________________________

____________________________________

____________________ ____________________

**Signed (optional)** **Building/ Department (optional)**

**Your cooperation in this survey is appreciated as we strive to improve our customer service.**

**Figure 3.17** *(Continued)*

- *Customer contact employees.* Train employees who have daily contact with customers to "close the loop" and provide information that customers should know. These employees are the first line of contact with these customers and know when something is not right. They can serve as an early warning system to alleviate potential problems. Additionally, a periodic group meeting with all customer contact employees can be instructional for them, through the sharing of ideas, and can also alleviate customer concerns.
- *Reports.* Distribute an annual report to customers. This lets them know what, and how well, the facility department is doing in meeting their expectations.

- *Periodic visits.* On an occasional basis the facility manager as well as all supervisors, and the customer service section, should visit customers in their work areas or send them electronic mail to determine how the customer is doing. This demonstrates genuine interest and care for the service being provided them. It helps to cement credibility and cultivate rapport.

## Quality Control Plan

The facility department (FD) quality control plan (QCP) should conform in every respect to corporate policies and procedures. A priority and an important management tool used to ensure maintenance of high-quality standards is a sustained quality control effort. The FD's QCP should ensure the customers' (tenants, building occupants, etc.) satisfaction with services received, work that is properly done, adherence to work performance schedules, and disciplined cost control. The QCP should be based on procedures that include an inspection system, administrative procedures for recording and tracking deficiency corrections, and timely record keeping and reporting.

### Specific QCP objectives

These objectives should ensure that the mechanical, electrical, plumbing, structural, energy management control systems, custodial and related services, structural alterations and improvements, and preventive and predictive maintenance work, and materials and parts utilized for the operation, repair, and modification of the facility will be of the best quality. This will include but not be limited to:

- Detecting and correcting existing or potential deficiencies
- Ensuring that FD personnel correct deficiencies right the first time
- Ensuring that employees know the expected work quality standards and are qualified to achieve them

- Ensuring that work performed is within established corporate acceptance quality levels (AQL)
- Conducting ongoing corporate trend analyses to ensure uniform quality and reliability
- Providing accurate QCP records and timely feedback reporting

The facility manager has ultimate responsibility for the QCP program design, development, implementation, and execution monitoring. Usually the FD staff size does not justify the cost of a separately staffed quality control inspector. Rather, the onsite QC program should link the QC efforts of FD supervisors, mechanics, and contractors; corporate managers; and FD professional and technical personnel.

Specific objectives of the QCP should include clear definition of organizational and individual responsibility. The goal is to involve every employee in the pursuit of work performance excellence. The QCP will be based on three separate undertakings: an overlapping system of inspections by FD supervisors; procedures for identifying, tracking, and correcting deficiencies; and administrative procedures for tracking, recording, and analyzing inspection results and determining trends in the quality of work performed. The following paragraphs describe an overall QCP that will meet all criteria and provide the best result-the inspection system; the checklists, forms, and other tools used; and the techniques used to track and correct deficient work. The final focus is on the overall QCP objectives.

### Inspection system

The FD staff should develop a detailed set of QC procedures, checklists, and guidelines for each provided service. The facility manager should then institute a series of interlocking QC inspections to be conducted by personnel on several organizational levels. These will include the facility manager, chief engineer, operating engineers, maintenance mechanics, cognizant corporate personnel, and contractor account managers.

### Inspections by FD personnel

The facility manager, chief engineer, operating engineers, and maintenance mechanics should inspect completed work, including service calls and maintenance and repair work orders. The chief engineer, operating engineers, and maintenance mechanics should tour the facility each day, conducting QC inspections informally. When their daily QC inspections are completed, covering their QC activities for the day, their reports should summarize all deficiencies noted during the QC inspections and provide recommended corrective actions. The facility manager should review the QC reports and forward them to the FD administrative assistant for processing. All daily submitted QC reports should be maintained in a hard-copy file for review by corporate managers.

Discrepancies noted during QC inspections should be keyed into the FD's computerized maintenance management system (CMMS, specifically into the QC module) initially for tracking purposes and into the service calls and maintenance and repair work order modules ultimately for corrective actions based on the deficiency's dollar value. Using these modules, deficiencies recorded (during QC inspections) will be tracked by the facility manager, chief engineer, operating engineers, and maintenance mechanics until corrective action is taken. Each deficiency will be assigned by the chief engineer to the appropriate mechanic (or subcontractor) for corrective actions.

Once corrected, the CMMS modules databases will be updated by the administrative assistant to include cognizant information and data, i.e., labor hours used, materials and parts used, equipment used, date completed, performing mechanic, etc. The facility manager will routinely assess pending workloads to plan for and assign the resources needed to correct the noted QC deficiencies. This action will be taken to prevent QC deficiencies corrections from being overlooked. Figure 3.18 illustrates the flow of QC inspection data in the CMMS from inception and discovery through to posting as completed jobs data bits ("cradle to grave" follow-up).

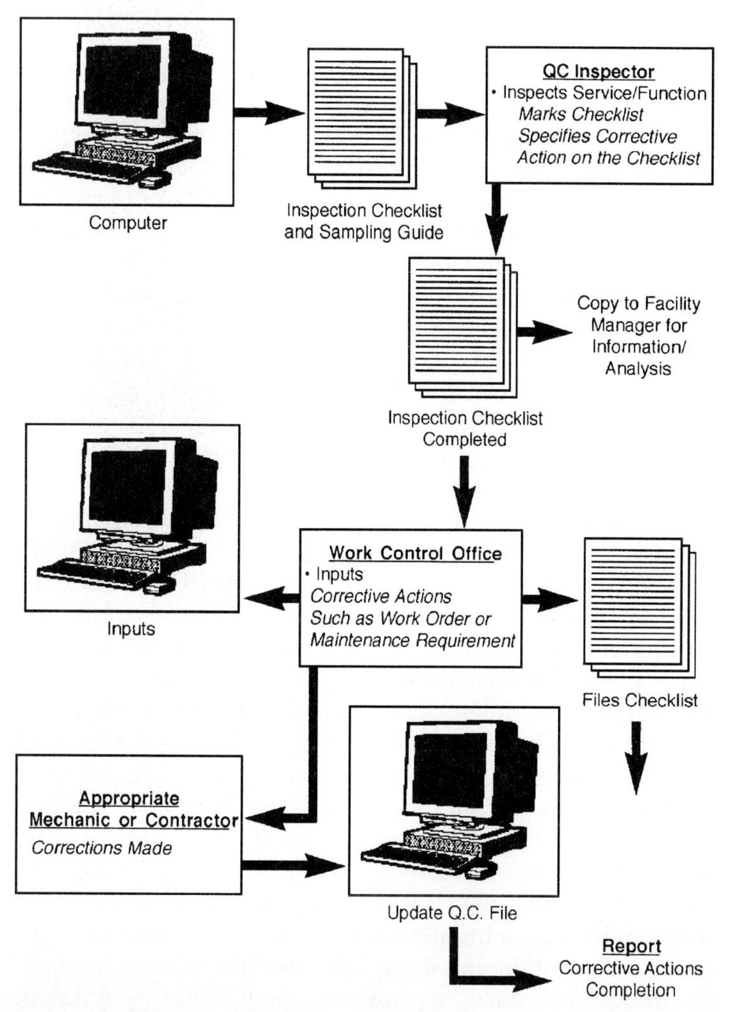


**Figure 3.18** Flow of quality control inspections.

## Identifying and tracking substandard performance

The QCP should provide a system of overlapping staff, corporate professional and technical personnel and managers, and contract managers to identify quality problems. Several techniques and tools should be used to assist these QC inspectors.

The primary tool should be facilities- and equipment-specific inspection checklists to be developed by technical experts. Another QC performance evaluation tool is the completed preventive maintenance work ticket form. Still other tools include visual observations, service call reports, and work order data. In general, for QC purposes, items inspected should include PM work, service calls and emergency callbacks, work orders issued and completed, tenant comments, tours and watch reports and logs, energy management control system reports and data, and historical repair records.

If the QCP is to be truly successful, it cannot be a "top-down" initiative only. Non-supervisory personnel must have a role in the process. Accordingly, maintenance mechanics, operating engineers, and other FD staff should be continually alert to note quality problems as they perform their duties. For example, mechanics often see poorly performed PM or repair work while engaged in preventive maintenance inspections on the same equipment item or one located nearby. When observed, deficiencies should be noted on the work order form. That information should then be annotated by the mechanic on the PM work ticket and keyed into the CMMS's QC module by the administrative assistant. The noted QC deficiencies should then be scheduled for corrective action by the facility manager.

Tour inspection reports are another practical means of identifying substandard maintenance work performance. Tour reports should be routinely reviewed by the facility manager, chief engineer, and operating engineers. QC problems should be processed as noted immediately above. Inspection reports and corrective actions taken should be maintained on file for further review, as required.

In addition to facility inspections, corporate managers and corporate professional and technical staff should review preventive maintenance work orders, service calls, and work order logs during their QC inspections. The priority given to a QC repair or deficiency observed should move higher the longer it remains on the service call log, either a manual or computerized log. The service call log is an important QC data source. The log should be reviewed daily by the chief engineer and operating engineers and reviewed periodically by the facility manager. Essentially, technical personnel should evaluate chronic calls, calls for the same problem, the number of calls received over time, etc.

### Correcting deficiencies

Tracking deficiencies from detection to correction is an important QC process. Deficiencies reported or identified by QC inspections should ultimately result in corrective action work orders. Like any work order, those documenting QC deficiencies should be assigned a work performance priority based on the criticality of work to be performed and then tracked through the company's CMMS. Such work orders should be identified as a QC correction. A computerized suspense file should be maintained for all QC corrective actions. This suspense file should be updated as the work is completed.

In addition to correcting specific problems reported (or otherwise identified), it is essential to identify casual factors leading to the deficient work. If not identified and addressed, casual factors lead to a cycle of repeated deficiencies and corrections. This trend drains productivity, job resources, and staff morale. Several strategies can be used to avoid this problem. For example, copies of quality control reports can be cycled from the FD back to the corporate engineering office. Data from these reports should then be extracted, input to a computerized database, and analyzed. These reviews yield trends or "root cause" analyses to be used by corporate staff engineers to detect the source of trends in work quality. When quality trends in a specific phase of work, area of the building, equipment, or system can be identified, determining the source of that trend becomes easier. Once the source is known, long-term corrective measures can be taken.

While such analyses are used to identify specific sources for negative (or positive) trends in work quality, various standard measures can be employed to proactively ensure that tasks are performed properly. Examples include:

- Writing quality standards into employee job descriptions
- Group training or one-on-one employee and supervisor training

- Observing employees throughout tasks performance
- Roundtable discussions with employees to elicit ideas that will improve work methods and resolve persistent quality problems

### Total Quality Management (TQM) program specifically for O&M Procedures

An FD TQM program, in concert with an organization's TQM program should be instituted for the purpose of achieving effective, efficient, and cost-effective O&M support efforts. The program should be designed to accomplish the following three important objectives:

1. Improving service quality by focusing on *how* work gets done, rather than *what* work gets done (i.e., methods as opposed to results).
2. Improving service without adding to staff or cost.
3. Establishing TQM as a continuing process in the FD.

To accomplish these objectives, several TQM methods should be used. These include but are not limited to the following examples:

- Instructing supervisors in alternative or new methods of performing their jobs.
- Having supervisors listen to facility tenants effectively, ensuring that the provided services meet their needs and are efficiently delivered.
- Permitting nonsupervisory personnel to improve work performance by contributing their ideas and giving them a stake in the success of those ideas.
- Basing operational decisions on data, not guesswork.
- Grouping in sequence all the tasks directed at accomplishing one particular outcome.
- Using a TQM team as a means of pooling skills, talents, knowledge, and ideas; also to obtain differing perspectives on a task.

The approach to using TQM principles should be symbolized by the relationship between work quality, a "scientific" or disciplined approach to studying the process, and advantages gained by a team approach, which is a shared objective. Working in tandem, these elements can be very effective and sustainable. The first TQM objective cited above focused on process, rather than results. It bears further discussion. Essentially, a process emerges by grouping in sequence all tasks directed at aparticular outcome, for example, the service call process. The *service call process* entails every action from call reception through job completion and upload of job data to the CMMS service call module. In evaluating service calls processing, TQM methodology should consider factors such as the following:

- Overall lead time to perform each action in the process
- Quality of work for each task performed
- Methods of reducing interference with tenant activities
- Response time
- Tenant satisfaction with work performed

Another important characteristic of TQM is the use of teams.

Rarely does one person have enough knowledge to understand all phases of a job or process. Therefore, major gains in quality and productivity often result from the use of teams, a group of process performing personnel pooling their skills, talents, and knowledge. With proper training, TQM project teams should be able to tackle complex and chronic problems and derive effective, permanent solutions.

Working as a team has another distinct advantage over solo effort: the mutual support that arises between team members. The synergy that comes from people working together productively on an important project is usually enough to sustain the enthusiasm and support, even through a difficult time. Project teams should be formed, over a period of time, to study all processes involved in FD service delivery, i.e., preventive and predictive maintenance, service calls, repairs, inventory control, warranty administration, energy management control system, etc. TQM teams should be organized by drawing personnel from the following sources:

- Corporate managerial, professional, and technical staff
- Facility Department staff
- Tenant representatives

TQM teams should be formed, one at a time, in a manner least disruptive to organization operations and activities.

# NOTES

Chapter

# 4

# Preventive and Predictive Maintenance Functions

## PREVENTIVE MAINTENANCE PROCEDURES

### Introduction

Properly conducted preventive maintenance procedures (PMP) reduce overall operating costs and aid in achieving facility department effectiveness and improved safety practices, thus assuring the continued preservation, usefulness, and performance of equipment and systems. These benefits are the primary justification for implementation of sound PMP. Preventive maintenance (PM) should not be considered a separate matter, but as an integral part of the total continuum of facility operations and maintenance.

#### Purpose of PM

PM carries the connotation that some action must be taken now to prevent a more serious problem at a later date. PM, as the name implies, means taking preventive actions now to eliminate the need for taking radical actions sometimes in the future.

### PM practice

PM practice means different things to different people. To most facility department technical and mechanical personnel, PM means inspections, adjustments, lubrication, parts replacements, and major overhauls. The term is broadly used, but it essentially represents a philosophy of providing care that will protect and maintain the essential operating quality of the equipment, system, or facility.

### Size of PM activity

The size of the PM activity is immaterial. To an electrical engineer, PM may mean the proper choice and setting of relays in controls to avoid unnecessary downtime. To a mechanical engineer, it may mean a complete overhaul of a pump or compressor.

### Definition of PM

To promote a clear understanding of the procedures, a simple, practical, and acceptable definition of PM should be established. First, to be considered within the realm of PM, an action must prevent some form of future deterioration or breakdown. The key element of PM consists of actions taken in the near term to prevent future breakdowns. PM, then, consists of periodic inspections, or checking, of existing facilities to uncover conditions leading to breakdowns or harmful depreciation of equipment and systems, and the correction of these conditions while they are still in the minor stage, including the preservation, lubrication, replacement of filters and worn parts, and other actions to prevent breakdowns or malfunctions.

### PM efficiency

PM ultimately reduces the corrective maintenance work load. As PM takes over, the corrective workload is shifted from when you *have* to do it, to when you *want* to do it. Thus, the work can be done more efficiently, at lower cost, and on schedule, with a substantially decreased likelihood of interfering with an organizations' program operations.

## A Rationally Planned PM Program Includes the Following Components

### Engineering input

The most valuable engineering input is in the original equipment, systems, and buildings design. PM will be facilitated if provisions for its performance are incorporated into the design phase. Designers often overlook the need for maintenance, or sacrifice the maintenance aspect in order to emphasize aesthetics or reduce initial costs. Later, the Facility Department (FD) must modify equipment and/or systems to provide workable equipment and systems that can be maintained at a reasonable cost. Obviously, such modifications will be costly (such as permanent platforms for accessibility rather than scaffolding).

Some of the responsibility for the frequency with which design and construction shortcomings are encountered must lie with the facility manager. In industry, in general, there has not been adequate promotion for better visibility for design flaws during the initial design phases of a new construction or alteration project to enable good facility department review and approval.

Inadequate operations and maintenance considerations in engineering design costs industry millions of dollars in unnecessary maintenance costs. The educational process must be more vigorously promoted by facility managers. Presenting proper and cognizant historical facts can do much to improve the situation.

Facility designers with good operations and maintenance understanding and background have contributed greatly to overall reduction of operations and maintenance costs. Sealed motors, modular components, and built-in diagnostic test units are a few of the innovations that have made maintenance an easier task. As the cost for operations and maintenance becomes increasingly significant, facility designers must produce more such innovations.

### Analysis of maintenance needs

When equipment and systems are installed, there is usually no index available for maintenance needs, except manufacturers' recommendations. Further details are acquired during day-to-day operation and maintenance of the equipment and systems. It is advisable to record these details. Analyzing these experiences by using simple measures of statistics, or probability, will give a sound basis for *what* must be done, *when* it must be done, and *to what extent* it must be accomplished.

### Proper use of systems and equipment

A PM program should include means for ensuring that equipment and systems are operated properly. This is accomplished by gaining cooperation of operations personnel. It has been determined that approximately 25 percent of all equipment and systems breakdowns are due to improper operations.

## Basic Controls

A well-engineered PM program, contrasted to the traditional "firefighting approach" requires skilled personnel; proper tools, materials, and supplies; comprehensive scheduling; and accurate record keeping and reporting. The last two activities, when properly established, become routine administrative tasks. One of the most efficient means available today for sorting, selecting, filing, and recording data is use of a *Computerized Maintenance Management System* (*CMMS*), software module for PM control.

## Field Inspection Procedures

Service maintenance manuals issued by equipment and system manufacturers are one of the best sources to facilitate the initial set-up of field inspection procedures. They are invaluable guides as to *what* and *when* to inspect, as well as *how* to service and maintain the equipment. The manuals usually contain checklists which itemize all points to be checked and maintained on the equipment and systems, thus providing a uniform and complete inspection. These manuals, coupled with experience, and engineering judgment, will normally provide sufficient information and data to ensure that used procedures are adequate. However, established procedures need continual refining and updating as experience dictates. During the process of revising manufacturers' recommendations to suit local facility conditions, it is safe to take the most stringent procedures selected from manufacturer's recommendations, engineering judgment, and current FD procedures to ensure optimum results.

No ready-made guide is available on optimum frequencies to perform PM inspections. Engineering analyses of operating cycles, failure history, and maintenance needs will be of invaluable assistance in determining and developing these frequencies. In making these determinations, consideration should be given to the following:

- Age, condition, and value (dollar value and value to the mission) of the equipment or system
- Severity of service
- Safety requirements
- Hours of operation
- Susceptibility to wear, exposure to dirt, friction, fatigue, stress, corrosion
- Susceptibility to damage, vibration, overloading, and abuse

PM performance frequencies should be periodically reviewed and adjusted. New equipment and systems must be checked frequently until they are "broken in." PM inspectors should be required to indicate recommended increases, or decreases, in the PM frequencies, using criteria based on repairs experience and statistics. Indications of the need to adjust PM frequencies include:

- *No repairs required:* Overmaintenance is indicated.
- *Frequent repairs required:* Inspectors are not getting to the trouble cause, or engineering assistance may be required.

#### Feedback, properly supplied and utilized

Feedback from supervisors, foremen, and mechanics is essential for the PMP to be successful. The first, and perhaps one of the most important, item needed is *man-hours used* and *material items and costs* expended. The second item needed is records of *completion* of assigned work tasks on the equipment and systems, together with a supervisor's verifications, to ensure a measure of technician productivity, work quality, and customer satisfaction. The third item needed is descriptions of delays encountered due to lack of transportation, safety permits, materials, tools, supplies, job sites availability, and outage scheduling.

Good engineering judgment should be utilized in executing the PMP. Techniques such as *Mean Time Between Failures* (*MTBF*) analyses, technical data accumulation and analyses, and work methods analyses should also be used. Written two-way communications between facility engineering supervisors, operating engineers, and craft mechanics should be used to resolve encountered problems.

#### Labor standards applied to PM

Initially, labor standard times should be used as a basis for estimating labor hour requirements. *Engineered Performance Standards* (*EPS*) are a set of labor time standards that have proven to be useful to many public and private sector organizations. However, as historical data on actual tasks performed are accumulated and then compared with task times estimates, it is possible to refine the EPS labor time standards to an overall accuracy of approximately ±5 percent. The identification of man hours required for the performance of equipment and systems PM would normally be developed by cognizant facility department work center craft and trade foremen in conjunction with the work control manager.

The application of labor time standards to repetitive type work, such as PM, is highly desirable. Analysis and interpolation of these data for similar, but not identical, jobs are excellent guides for PM planning. This approach, coupled with engineering judgment and past experience of qualified personnel, provides the starting point for applying labor hour standard times. Continual evaluation will further assist in refinement of the labor hour standard times.

### How PM relates to other routine maintenance

The PMP should be integrated with regular facility maintenance and engineering programs. Supervisors, work planners, and work schedulers who handle PM assignments should be the same as those who direct, plan, and schedule regular maintenance work. Mechanics should be rotated periodically so that as many as possible acquire familiarity with the derived procedures.

### Additional PM program techniques

After the basic program has been installed and proven economically beneficial, there are other improvement techniques that can be applied. These include the following:

- Protective methods
- Coatings—cathodic and plastic
- Alarms—vibration mounts
- Craftsmen training
- Standard practice manuals
- Materials research

## Facilities Inspection and Maintenance Program

As facilities become more complex, and more resources are required to provide and maintain needed facilities, the question is how can FDs keep up with operations and maintenance requirements?

One way would be to hire more operating engineers and mechanics to take care of both the new and existing requirements. But since needed additional resources are usually not available, this is not a viable option. Any available new resources are usually earmarked for upgrading facilities, equipment, and systems. But this is the way it should be, for any new equipment and systems generally make facilities operations more efficient and cost effective. So what is the answer? How can the facility manager do more with less? A new direction is needed to face this dilemma if you are going to keep up.

One approach is to use a set of programs known as the *Facilities Inspection and Maintenance Program* (*FIMP*). Although the success or failure of an FD is hard to ascertain, and frequently must be based on empirical evidence, there is evidence that FIMP use has been successful and has helped to meet the changing needs of organizations, without placing undo strain on the organization's budget. Generally, the results will be evident in:

- Staff and payroll decreases
- Facilities in better shape
- Square footage requirements, per employee, increased with no ill effect
- Contract costs remain level

If prevention, or decrease, in deferred maintenance projects is a criteria for a successful FD, then FIMP has proven to be a success in many organizations. How will you know if facilities users (customers) are more satisfied? Note the following indications:

- The FD's work reception telephone is not constantly ringing with requested work requirements or emergencies.
- Irate telephone calls are very rarely received in the Department because the majority of work is done in a timely fashion, and much is done before a customer could even notice that something needed to be done.

- Even though more area is being maintained with fewer people, emergency and service repair tickets volume has not increased.

These results can be achieved using the FIMP programs. However, other alternative solutions should be evaluated also before deciding to use FIMP. Some alternatives are as follows:

*Computers.* Computers are good management tools, but you do need the necessary data bases to readily determine the required operations and maintenance effort. Computers may help with keeping track of what was done; but, a method is needed to provide direction to the required operations and maintenance effort.

*Reduction in overall maintenance effort.* Reducing the overall maintenance effort, accepting the fact that you cannot get to all requirements anyway, is not the answer because lowered maintenance efforts leads to a downward spiral for a facility's level of maintenance that does not bottom out as equipment and systems age, increase in size, and become more complicated.

*Doing only what needs to be done.* Another approach is doing only what needs to be done and skipping the rest. This takes care of essential equipment and systems, and keeps the facility operating; but invariably the areas where maintenance levels are reduced are the areas that impact on an organization's operations and cause problems.

*Plan some maintenance and fire-fight the remainder.* This could work, but performing maintenance this way means some customers could learn the FD's maintenance management system and get all their work done while others would get almost no work done as they did not know how to use the system to their advantage.

From all the reasons listed above, it has been deduced that there was a better way. The better way is actually the old ways which are incorporated into FIMP.

Major areas of improvement will result from very minor changes and constant positive reinforcement of old proven procedures. Any FD must constantly inspect everything to keep abreast of current conditions. Do not hire special inspectors, rather it is better if you use existing personnel. A carpenter or locksmith knows more about maintenance and repair requirements of door hardware than a facility manager normally would, and a building manager would know more than a user. One particular area where you can achieve considerable success is with regular housekeeping and custodial inspections.

These inspections, conducted by FD personnel, accompanied by custodial and housekeeping supervisors and building occupants, will spot flaws in custodial and housekeeping routines,136 but ultimately will accomplish much more. The most obvious results will be cleaner, neater buildings. But most importantly, in addition, regular inspections show the "flag" by having FD personnel, housekeeping personnel, and building occupants dealing with one another. This results in better rapport and better understanding of problems associated with custodial maintenance. Housekeeping inspections are not covered by FIMP, but are essential to its success due to the ease of spotting maintenance requirements when buildings are being cleaned and occupants become accustomed to dealing with FD personnel. After rediscovering the correct maintenance methods, the next step is getting organized to complete the job.

You begin by determining goals. The goal for maintaining life safety equipment is easy. Any condition other than 100 percent is not acceptable. Also, with comparatively reduced financial resources, you cannot afford any emergency work which would have to be deferred. The chosen method would have to allow for an early warning of developing problems. Any system chosen needs to allow for tight manpower scheduling and for rapid labor force movement to concentrate on specific areas, because the window of time availability in some buildings is often quite narrow. Another requirement is to combine usage and visibility in a coherent system; invariably some areas are heavily used and few people see them (kitchens), while some areas are not used but everyone sees them (elevator lobbies).

Considering these constraints and objectives, you are able to commence your implementation planning. To keep up with the maintenance requirements, a good place to start is where all concerned can agree on what is needed to determine necessities. As mentioned previously, maintenance requirements for life safety equipment is easy to determine; sprinklers, fire alarms, fire door components, smoke and heat detectors, regular and emergency electrical systems etc. must be kept in like-new condition. Other building components need greater or smaller amounts of attention. Determining the correct level of maintenance for any one system requires having a complete, current inventory on hand; research into manufacturers recommendations; estimates of required man hours to perform the work; and many educated judgment calls. In this manner, work through each facility's systems and equipment; and then combine the results into a single FIMP for scheduling purposes.

This system of integrated programs provides management with historical files for each building. As data are accumulated in each program file for each building, it becomes easier to assess needs and pinpoint areas where problems are developing. Deferred maintenance is thus avoided and manpower is easily adjusted to accurately develop current needs.

FIMP can be originally implemented by drawing together existing programs for centralization. Existing programs form the backbone for FIMP development. One early program to be evaluated is roof inspections. Roof inspections can best be performed by contractor personnel, rather than in-house personnel, due to the specialty craft trade requirements. It is absolutely essential that roofs be well maintained to ensure water tight building envelopes. Ideally roofs should be inspected annually using an in-house inspector who:

- Can perform minor maintenance on-the-spot
- Knows roofing maintenance procedures in detail
- Can supervise outside contractors to ensure quality workmanship is provided

There should not be too much redundancy in roof inspections. Roof inspections should include nondestructive tests with a nuclear device and infrared spot checks, taking regular core samples of roofing membranes, and twice-a-year thorough visual checks. It is essential to remember that it takes only a few weeks for a roof to go from being maintainable to requiring total replacement. Other programs can be added as required. It is easy to identify areas requiring routine maintenance. Most programs derive from customer needs or complaints. Other programs will be designed to meet specific needs, such as sprinklers, for insurance purposes. As a need for a particular equipment or system program becomes evident, it is necessary to take the following actions:

- Calculate the frequencies needed for preventive maintenance visits
- Conduct an inventory of all similar equipment and systems, including the recording of name plate data, locations, and current levels of maintenance
- Develop needed craft manpower requirements to conduct the preventive maintenance visits, including labor hour times for site visits for work, travel, preparation, and delays encountered
- Develop checklists to control each program, using industry standards, past experience, and engineering judgment

Each program, and its controlling documents, should be designed for flexibility to meet FD, user, and organization needs. For example, manufacturers recommend that emergency diesel power systems should not be started and run without load, but to ensure users that they will start, you can start them once a week anyway; and then once a month perform a full load test.

A third reason to start FIMP programs is for FD convenience. An example of this type program is kitchen maintenance. It is easier for the FD to allow the kitchen users (they are the experts) to be responsible for maintenance; but, it is more efficient and effective to have the FD schedule the maintenance on a regular, routine basis. Otherwise, craftsmen would be responding to service and emergency calls, and repairing one item at a time.

Programs should be divided into three categories during the development phase:

1. Inspections are programs used to define and identify breakdowns in major equipment or systems such as roofs or structural envelopes, or to look for particular items, such as asbestos during remodeling projects. When a needed repair is identified, a work order will be issued and prioritized. These repairs, controlled by the work order, are either performed according to priority or forwarded to the facility manager for approval, funding, and designation of the performance method, i.e., in-house or by contract.
2. *PM* programs are also inspections, but with *predetermined maintenance* and *benchmark tests* included. The maintenance and tests needed for a particular equipment or system are defined and controlled by checklists. Any discovered problems are usually corrected as emergencies.
3. *Find/Fix* are programs that use scheduled teams of craftsmen to systematically inspect for, and then perform, contingency maintenance on a particular system or equipment, or a complete building. During a Find/Fix procedure virtually any problem discovered is repaired immediately.

In each of these categories the individual programs are broad in scope, yet detailed in application. Program examples are preventive maintenance of major switchgear and Find/Fix for buildings' interiors.

Maintenance of major switchgear is critical for the usefulness of a building for obvious reasons. How does an FD ensure that all switchgear components are in design ranges when the work must be performed by specialty contractors not familiar to FD personnel? To manage this type of program, the FD must have complete inventories of all equipment and systems in each building, and a complete set of specifications. These specifications are available from many sources, including the requirements from the appropriate governmental (federal, state, county, and/or city) jurisdiction. The specialty contractor should be tasked to provide full reports, based on accurate guidelines, listing the test results on each piece of switchgear equipment. The report should also list further maintenance requirements needed. This report will be the FD's control over work performance, and the basis for approving invoice payment.

Another program example is the Find/Fix series. It should be the FIMP cornerstone for the FD. Normally this program will use approximately one-half of the routinely scheduled man hours, control the interior maintenance of buildings, and prevent the deterioration of offices, laboratories, conference rooms, etc.

The normal maintenance management approach is to inspect building interiors, and then issue corrective maintenance work orders as needed. This method focused maintenance on necessary areas; but the cost of inspectors, long lead time to complete the repairs, and attendant paperwork made the process economically infeasible. Find/Fix enables the FD to deploy craftsmen to find the needed repairs and fix them on the spot.

The Find/Fix method ensures that large costly repairs would not result from unnoticed small repairs. The central theme is similar to the method used by homeowners: being observant, making small repairs as needed, or ensuring that someone else makes the repairs. The concept is simple in design, but implementation on a large scale, with large crews, requires much supervision and training, and is difficult and time consuming. A workable program schedule development requires the following elements: needed materials inventories; crew sizes calculations; average production in square footage per man-hour; and a schedule of availability of buildings for the program. After the program is documented and scheduled, checklists that include each item thought to need regular attention should be designed and developed for each participating craft.

The keys to successful maintenance practices are:

1. Go back to basics such as Find/Fix inspections.

2. Forget computers until your programs are in place and well integrated.
3. Inspect to the point of redundancy, if at all possible make repairs on the spot, and stick to what you do well—minor maintenance and repairs and not general contracting work.
4. Use remodeling work to fill in the low spots, use contractors to level off the high spots, and stick to maintenance.
5. Use of FIMP will allow you to *routinely* schedule approximately 50 percent of available man-hours, and find and direct another 30 percent through inspections.

The use of FIMP scheduling procedures allows you to plan the use of the remaining 20 percent almost year-round. FIMP takes a great deal of planning and top management support to make it work; *but it does work.*

## Facilities Inspection and Maintenance Program Specifics

*Envelopes.* Annually, the structural inspector should perform a complete structural envelope inspection using a check list and a set of instructions.

*Roofs.* Annually, a roof inspector should perform a complete roof inspection on each building. The roof inspector should submit inspection forms, complete with findings and repair cost estimates to the facility manager for approval and subsequent actions.

*Moisture meter.* Every other year moisture meter tests should be performed on each major flat roof. These tests are performed to establish benchmark data, and to pinpoint problems, before breakdowns.

*Sprinkler inspections.* Quarterly, in-house personnel should perform sprinkler inspections, which entail complete checks of flow and tamper switches, alarms, etc. The discovered problems should be reported, and processed as "emergencies" in the maintenance management system.

*Relamping.* Performed, as required on a case-by-case basis, using in-house personnel.

*Switch boards.* Every three years, contract personnel perform switchboard inspections on systems over 1000 AMPS. The inspections should be controlled by FD documents.

*Motor control centers.* Performed every three years, by contract personnel, testing and maintaining contacts and heat sensors. The testing and maintaining should be controlled by FD documents.

*Emergency generators.* In-house personnel, directed and controlled by checklists, inspect and repair building interiors.

*Find/Fix on diesel generators.* Twice a year, in-house personnel, controlled by checklists, perform the following listed tasks on the diesel generators:

- *Weekly.* Check all gauges, connections, batteries, and fluids.
- *Monthly.* Perform a complete operational check under load.
- *Emergency generators.* Weekly, in-house personnel, controlled by checklists, perform complete system checks on gasoline generators.

*Emergency battery packs.* Monthly, in-house personnel, controlled by checklists, test and check for voltage and specific gravity.

*Trash chutes.* Quarterly, in-house personnel, controlled by checklists, test and repair trash chutes.

*Fire alarm systems.* Annually, in-house personnel, controlled by FD documents, test and repair fire alarm systems by activating each station; and monitoring alarms, bells, and panels. Necessary repairs are performed as follow-ups on an emergency basis.

*Elevators.* Weekly, bi-weekly, or monthly, contract personnel, controlled by FD documents and specifications, perform maintenance and/or repairs as needed.

*Burglar alarm systems.* Monthly, in-house personnel test each burglar alarm system and perform repairs as needed.

*Fume hoods.* Annually, contractor personnel test and certify all fume hoods and exhaust systems for proper velocity and pressure.

*Smoke detectors.* Semiannually, in-house personnel, controlled by checklists, inspect, test, or repair smoke detectors as required.

*Sprinkler systems.* Monthly, in-house personnel, controlled by checklists, perform preventive maintenance on wet systems, and weekly (twice a week in freezing weather) on dry systems.

*Irrigation systems.* Twice a year, in-house personnel, controlled by checklists, inspect and repair irrigation systems. Systems should be charged in the Spring and made ready for the watering season, and winterized in the Fall.

*Halon systems.* Annually, contract personnel with in-house assistance maintain Halon systems.

*Painting.* On a recurring five-year basis, paint all buildings.

*Clocks.* In-house personnel reset clocks in April and November to reflect the changes from EST to DST.

*Site drains.* Quarterly, in-house personnel inspect and clean, as necessary, site drains.

*Water heaters.* Annually, in-house personnel perform routine maintenance on large capacity gas fired water heaters.

*Uninterruptable power supplies.* As required, contract personnel maintain uninterruptable power supplies in appropriate facility areas and rooms.

*Find / Fix Building Interiors.* Three times a year, in-house personnel, controlled by checklists, inspect eye washes and safety showers for pressure and water volume.

*Lightning suppression systems.* Annually, in-house personnel, controlled by checklists, inspect lightning suppression systems for ground and continuity.

*Acid tank Find/Fix.* Twice a year, in-house personnel, controlled by checklists, inspect and repair acid neutralizing tanks.

## Equipment and Systems Maintenance Procedures

### Preventive maintenance job tasks

Formal written job tasks should be established for each type of equipment and system in the facility's PMP. The tasks should be written so as to ensure that each particular job task can be performed on more than one equipment or system item. Figures 4.1 through 4.50, inclusive, show typical task descriptions for a representative range of equipment and systems. Figures 4.1 through 4.50 are used with Georgetown University's approval.

Each item or step of the job to be performed is listed with a description of the work to be done. The description of the job is laid out in a stepwise manner and written in terms that the mechanic doing the work can understand. The complete task is packaged, given a suitable reference number, and filed and referenced in such a manner so as to be easily retrievable—normally part of the Computerized Maintenance Management System's PMP module.

### Personnel requirements

*Refrigerants.* Personnel who handle refrigerants containing chlorofluorocarbon refrigerants (CFC) must pass an EPA-approved examination, to achieve a level IV (Universal) certification.

*Energy Management Control System (EMCS).* Personnel who operate, maintain, and repair EMCS must complete the following: (1) a training course provided by the EMCS manufacturer with a certificate of completion, and (2) computer programming experience with EMCS.

*Craftsmen.* Journeymen trade specialists should be licensed and/or certified by the appropriate local governmental jurisdiction—city, county, or state.

*(Text continues on p. 201.)*

GEORGETOWN UNIVERSITY PREVENTIVE MAINTENANCE(GUIDE NUMBER) 1

EQUIPMENT: Electric Motors

MANUFACTURER: All

CATEGORY: Electrical FREQUENCY: Yearly

PREVENTIVE MAINTENANCE DESCRIPTION:

Conduct Ammeter reading of motor electrical load.

SPECIAL INSTRUCTIONS:

- Exercise special caution while working near energized circuits. Test is to be made on motors operating at full load.

PROCEDURE:

1. Conduct ammeter reading of motor electrical load:

   a. Open the cover door of the motor starter/controller unit or the motor disconnect switch (Actuate defeater catch to open).

   b. With the motor running, take ammeter readings of each of the three motor lines at the output terminals of the contacts.

   c. Observe the amperage load of the motor and compare it with the amperage load for that particular motor.

   d. Record readings on preventive maintenance work order.

TOOLS AND MATERIALS: Revision Date: 8/10/88
Standard hand tools, clamp on ammeter

**Figure 4.1** PM Guide Number 1—Electric motors. (*From Georgetown University.*)

GEORGETOWN UNIVERSITY PREVENTIVE MAINTENANCE(GUIDE NUMBER) 2

EQUIPMENT: Rotating AC Machinery

MANUFACTURER: All

CATEGORY: Test FREQUENCY: Yearly

**PREVENTIVE MAINTENANCE DESCRIPTION:**

Megger Insulation Resistance Test (All rotating machinery)

**SPECIAL INSTRUCTIONS:**
OSHA Standard 1910.213 -Open lock and tag motor disconnect switch and starter before starting work. Check the voltage of all incoming line terminals. Positively ascertain that the equipment is deenergized.

**PROCEDURE:**
Insulation resistance should be checked once a year. The test must be done under the same conditions each time in order for the readings to have comparison value, (motor temperature, operating time before test is made, etc.). The following formula will render the minimum resistance value for the equipment. With the stator windings at their normal operating temperature, measured at 500 volts d.c., one minute after motor has been turned off and ready to test.

$$\frac{\text{rated voltage of machine} + 1000}{1000} = \text{insulation resistance in megohms}$$

A. Use a thermometer to check the temperature of the motor. Do this at the top center of the motor housing. Record the temperature reading on the PM form.
B. Deenergize all power to the motor that is being tested.
C. Connect the megger ground lead to a good ground source and test the megger. Now proceed to check each motor lead one at a time. Record the insulation resistance of each test on the PM form.

*Note-In the event the readings do not meet the minimum operating standards as previously calculated you should report the deficiency to the shop foreman responsible for the repair or replacement of the motor.

**TOOLS AND MATERIALS:** Revision Date: 8/5/88
Voltage Tester, Safety tags and locks, Megger meter, Hand tools to remove covers, thermometer to read motor temperature.

**Figure 4.2** PM Guide Number 2—Rotating AC machinery. (*From Georgetown University.*)

GEORGETOWN UNIVERSITY PREVENTIVE MAINTENANCE(GUIDE NUMBER) 3

EQUIPMENT: Electric motors(all)

MANUFACTURER: All

CATEGORY: Cleaning FREQUENCY: Yearly

PREVENTIVE MAINTENANCE DESCRIPTION:

Cleaning and Drying Electric Motors

SPECIAL INSTRUCTIONS:
OSHA Standard 1910.213 -Open lock and tag motor disconnect switch and starter before starting work. Check the voltage of all incoming line terminals. Positively ascertain that the equipment is deenergized. Volatile Solvent cleaning agents are to be used in well ventilated areas only -following the manufacturer's recommendations.

PROCEDURE:
The motor exterior should be kept free of oil, dust, dirt, water, and chemicals. For fan cooled motors, it is particularly important to keep the air intake opening free of foreign material. Dirt buildup inside or outside of motors can cause overheating and shortened insulation life. A motor subjected to a 10°F temperature rise above the maximum operating temperature will cause premature motor failure.

Cleaning

Clean the motor inside and outside regularly. Use the following procedures, as they apply:

A. Wipe off external surfaces of motor. Use a petroleum solvent if necessary.
B. Remove dirt, dust, and other debris from ventilating air inlets on enclosed motors. Remove end-plate covering the motor fan. Brush and wipe the fan blade clean.
C. Clean open motors internally by blowing with clean, dry, compressed air at 40 psi.
D. When dirt and dust are solidly packed, or windings are coated with oil or greasy grime, disassemble the motor and clean with solvent. Use only high flash naphtha, mineral spirits, or stoddard solvent. Wipe with solvent dampened cloth or use suitable soft bristle brush. Do not saturate. Air dry stator for 24 hours. Then oven dry at 150°F. Be sure windings are thoroughly dry before reassembly.
E. Encapsulated stator windings may be rinsed or sprayed with solvent and immediately wiped dry with a cloth. These windings may also be cleaned with water and a fugitive detergent(ammonium oleate). Rinse with clean, clear water to remove all detergent. Hot water or low-pressure steam may also be used. Wipe excess water from metal surfaces and oven dry at 200°F.

Pg 1 of 2

**Figure 4.3** PM Guide Number 3—Electric motors (all). (*From Georgetown University.*)

GEORGETOWN UNIVERSITY PREVENTIVE MAINTENANCE(GUIDE NUMBER) 3
(continued)

Cleaning

F. Motors may be dried out by heat from a warm air oven, electric strip heaters, heat lamps, or by passing current through the windings. The temperature should not exceed 167°F. A forced circulation type oven rather than a radiant type oven is recommended. Radiant ovens tend to scorch and burn surfaces before the desirable dryness is reached.

When the motor is dried by passing alternating current through the windings with the bearing housings removed, the rotor must be centered in the stator core. Make certain that the air gap is uniform by wedging fiber strips in the lower portion of the gap. A controlled current of the same number of phases and the same or lower frequency is applied to the terminals.

Voltage should not exceed 10% of normal, and it should not cause more than 60% of normal full-load current to pass through the windings. A voltage 15% of normal may be applied after the insulation resistance has reached ½ of the minimum value for the insulation normal operating resistance. The temperature should not exceed 167°F during drying by this procedure.

G. After cleaning and drying windings check the insulation resistance with a megger test meter. See PM Guide 2 for procedure.

TOOLS AND MATERIALS: Revision Date:8/5/88
Air compressor, solvent cleaner, soft wire brush, rags for cleaning, vacuum, hand tools to remove covers

Pg 2 of 2

**Figure 4.3** *(Continued)*

GEORGETOWN UNIVERSITY PREVENTIVE MAINTENANCE(GUIDE NUMBER) 4

**EQUIPMENT:** Electric Motors Grease by Disassembly

**MANUFACTURER:** Louis Allis

**CATEGORY:** Lubrication **FREQUENCY:** 3 Years

**PREVENTIVE MAINTENANCE DESCRIPTION:**

Remove motor end shields and repack bearing. (Louis Allis)

**SPECIAL INSTRUCTIONS:**
OSHA Standard 1910.213- Open lock and tag motor disconnect switch and starter controller before starting work.

**PROCEDURE:**

1. Re-grease motor bearings by disassembly:

a. Motor must be disassembled to lubricate the bearings on this type of motor.Remove end shields from motor, clean grease cavity.

b. Refill grease cavity three quarters full with No. 2 Consistency grease with a -10° to 200°F Ambient.

Caution: Bearings and grease must be kept free of dirt.

**TOOLS AND MATERIALS:** Revision Date:11/7/88
Standard hand tools, grease, rags.

**Figure 4.4** PM Guide Number 4—Electric motors grease by disassembly. (*From Georgetown University.*)

GEORGETOWN UNIVERSITY PREVENTIVE MAINTENANCE(GUIDE NUMBER) 6

EQUIPMENT: Electric Motor, Ball Bearing, Flush Type

MANUFACTURER: G.E/Westinghouse/Baldor/Louis Allis

CATEGORY: Lubrication FREQUENCY: Yearly

PREVENTIVE MAINTENANCE DESCRIPTION:

Regrease single shielded motor bearings.

SPECIAL INSTRUCTIONS:
OSHA Standard 1910.213 Open lock and tag motor. Disconnect switch before starting work.

PROCEDURE:

A. Regrease single shielded motor bearings with the motor at normal operating temperature.

B. Wipe the lubrication fitting, bearing casing and grease relief plug clean at both ends of the motor housing to insure no dirt enters bearings.

C. Remove the relief plugs;if grease has hardened, probe a short distance into grease chamber to break up grease so it will be forced out by the new grease.

D. Insert new grease through the lube fitting with a low-pressure grease gun until new grease appears at the grease relief holes.

E. Energize and run the motor for ten minutes so the bearing can expel any excess grease. Probe the new grease from the relief hole to allow for expansion and replace plug.

F. Always wipe up expelled grease from around relief plug and filler fitting.

TOOLS AND MATERIALS: Revision Date:8/5/88
Standard hand tools, rags, grease gun.

**Figure 4.5** PM Guide Number 6—Electric motors, ball bearing, flush type. (*From Georgetown University.*)

**GEORGETOWN UNIVERSITY PREVENTIVE MAINTENANCE(GUIDE NUMBER)** 9

**EQUIPMENT:** Preheat / Reheat pumps / H.W circ. / CH.W Circ

**MANUFACTURER:**

**CATEGORY:** **FREQUENCY:**

**PREVENTIVE MAINTENANCE DESCRIPTION:**

Lubricate the pump motor bearings.

**SPECIAL INSTRUCTIONS:**

Clean any dirt buildup from air intake openings of motor and wipe motor clean.

**PROCEDURE:**

1. Taco Pump Motors:

   a. Add one teaspoon of SAE20 ND oil to each motor bearing.
   **Caution:** do not over-lubricate motor bearings.

2. For Thrush Vertical Mounted Pump Motors:

   a. Add approximately 1/4 ounce of NO2 -20° to 300°F Ambient grease to upper and lower bearings.
   **Caution:** do not over-grease

**TOOLS AND MATERIALS:** **Revision Date:** 11/01/88
Standard Hand Tools, Grease Gun with NO2 -20° to 300°F Ambient, Grease Oilcan with SAE 20 Non Detergent oil, 6 foot ladder.

**Figure 4.6** PM Guide Number 9—Preheat/reheat pumps/H.W. circulation/CH.W. circulation. (*From Georgetown University.*)

GEORGETOWN UNIVERSITY PREVENTIVE MAINTENANCE(GUIDE NUMBER) 15

EQUIPMENT: Coils and Fans

MANUFACTURER: All

CATEGORY: Cleaning FREQUENCY: Yearly

PREVENTIVE MAINTENANCE DESCRIPTION:

Coil and Fan Cleaning.

SPECIAL INSTRUCTIONS:

Never steam clean closed pressurized systems. Use eye protection at all times. Be familiar with chemical cleaning agents you are using and their hazards. Use gloves and protective clothing.

PROCEDURE:

Fan Blades, Condenser Coils, Chill and Hot Water Coils, Reheat Coils, Preheat Coils.

Where cleaning can be done with water and detergent and only minimum precautions are necessary to prevent the water damaging any surrounding areas this will be the procedure of first choice. In the event water cannot be used compressed air will be blown through the coils after brushing the coil face. This procedure will be repeated on the back of the coil followed by a thorough vacuuming of each side of the coil and compartments.

A steam cleaner is acceptable in lieu of water and detergent but in no case is steam cleaning to be used on coils containing an isolated fluid or gas. The resultant expansion from hot steam on the coil will produce explosive results and severe personal injury.

The complete removal of dust, oil, and grease is the objective. If, during the cleaning process, surfaces loose their protective coatings a suitable means to restore that protection should be applied as soon as the surface is dry.

TOOLS AND MATERIALS:
Gloves, Coveralls, Wet-vac, Hose, Air Compressor, Brushes, Rags, Eye Protection.

**Figure 4.7** PM Guide Number 15—Coils and fans. (*From Georgetown University.*)

GEORGETOWN UNIVERSITY PREVENTIVE MAINTENANCE(GUIDE NUMBER) 16

EQUIPMENT: Reheat Coils

MANUFACTURER: All

CATEGORY: Cleaning FREQUENCY: Yearly

PREVENTIVE MAINTENANCE DESCRIPTION:

Clean and Check Reheat Coil.

SPECIAL INSTRUCTIONS:

PROCEDURE:

1. Clean, Inspect, and Check Reheat Coils:
   a. Open access cover upstream of reheat coil and clean coil face of dirt and debris by brushing and vacuuming.

   b. Inspect coil for damage, bent fans, evidence of leaking or corrosion.

   c. Actuate valve with thermostat to ensure that it functions.

TOOLS AND MATERIALS: Revision Date: 8/10/88
Vacuum, Soft wire brush, Drop light, 9' ladder, hand tools to open covers.

**Figure 4.8** PM Guide Number 16—Reheat coils. (*From Georgetown University.*)

**GEORGETOWN UNIVERSITY PREVENTIVE MAINTENANCE(GUIDE NUMBER)** 17

**EQUIPMENT:** Condensing unit(For Walk-In Box)

**MANUFACTURER:** Punham-Bush/Tecumseh/Copeland/Carrier

**CATEGORY:** Refrigeration **FREQUENCY:** Yearly

**PREVENTIVE MAINTENANCE DESCRIPTION:**

1. Conduct annual maintenance of walk-in box condensing unit compressor.

---

**SPECIAL INSTRUCTIONS:**

Exercise caution when working on running equipment.

---

**PROCEDURE:**

1. Conduct annual maintenance of walk-in condensing unit:

   a. Check level of compressor oil.

   b. Flush the water regulating valve by inserting a screwdriver under the control valve arm and lifting the valve a couple of times.

   c. Check tightness of all flare nuts.

   d. Wipe down unit, inspect for any damage, oil or freon leaks.

   e. Tighten all electrical wire terminals on machine; inspect for burned or worn contact points. Check electrical wire terminals on starters for tightness.

   f. Lube motor only in accordance with the assigned PM guide.

   g. Check refrigeration charge with manifold set. Recharge if needed. Report leaks to supervisor and note if there is a leak on PM card.

   h. Ensure that all belt guards are in place when done.

---

**TOOLS AND MATERIALS:** Standard hand tools, AC manifold set, rags, grease gun, Freon for recharging.

**Figure 4.9** PM Guide Number 17—Condensing unit for walk-in box. (*From Georgetown University.*)

GEORGETOWN UNIVERSITY PREVENTIVE MAINTENANCE(GUIDE NUMBER) 19

**EQUIPMENT:** Air Dryer

**MANUFACTURER:** Hankison/Johnson

**CATEGORY:** Air dryers **FREQUENCY:** Yearly

**PREVENTIVE MAINTENANCE DESCRIPTION:**

1. Clean and check air dryer.

**SPECIAL INSTRUCTIONS:**

Unplug unit or lock out switch before starting work.

**PROCEDURE:**

1. Clean air dryer refrigeration unit:
   a. Turn off unit; remove access panel from cabinet.
   b. Clean dust and dirt from the inside of the cabinet, the condensing unit, and especially from the condenser coil.
   c. Wipe the fan blades clean and rotate fan by hand to check the condition of the fan bearings.
2. Clean and check filters:
   a. Check the oil filter if the media indicator has turned pink. Replace the desiccant filter cartridge.
   b. Automatic drain: operate the float valve manually during inspection. Clean automatic drain if desiccant filter is changed. Use soap and water only.
   c. Check by-pass valve to make sure all flows are through the condenser and oil filter.

**TOOLS AND MATERIALS:** Standard hand tools, strap wrench to open filters if needed, rags.

**Figure 4.10** PM Guide Number 19—Air dryer. (*From Georgetown University.*)

GEORGETOWN UNIVERSITY PREVENTIVE MAINTENANCE(GUIDE NUMBER) 20

EQUIPMENT: Humidifier

MANUFACTURER: All

CATEGORY: Heating and AC FREQUENCY: Yearly

PREVENTIVE MAINTENANCE DESCRIPTION:

Steam Humidifier inspection and cleaning

SPECIAL INSTRUCTIONS:

PROCEDURE:

Operate the control valve. Check the discharge from the manifold. If the unit is not working properly do the following.

A. Check control valve for dirt or foreign material. If necessary, clean.

B. Check the strainer screen. Remove and clean.

C. Check trap for dirt and foreign material, clean if necessary.

D. Reassemble and test unit.

TOOLS AND MATERIALS:
Two pipe wrenches, screw driver, wire brush, pressure bulb or other means to operate pneumatic valve.

**Figure 4.11** PM Guide Number 20—Humidifier. (*From Georgetown University.*)

GEORGETOWN UNIVERSITY PREVENTIVE MAINTENANCE(GUIDE NUMBER) 21

EQUIPMENT: Fans

MANUFACTURER: All

CATEGORY: Cleaning FREQUENCY: Yearly

PREVENTIVE MAINTENANCE DESCRIPTION:

Corrosion control (Fans and Fan Housings).

SPECIAL INSTRUCTIONS:

OSHA Standard 1910.213 -Open lock and tag motor disconnect switch before starting work.

PROCEDURE:

When the fan is equipped with access panels remove them before starting inspection.

A. Inspect the fan casing and blades for corrosion and any accumulation of dirt.

B. Use a dry rag to remove dirt.

C. Use Varsol to remove dirt that does not come off by wiping with a dry rag.

D. Remove corrosion with a wire brush and paint the area with spray primer paint. Fan shaft corrosion should be removed with emery cloth and the shaft recoated with a stripable corrosion resistant coating.

TOOLS AND MATERIALS: Revision Date: 8/10/88
Rags, wire brush, drop light, primer paint, scrapper.

**Figure 4.12** PM Guide Number 21—Fans. (*From Georgetown University.*)

GEORGETOWN UNIVERSITY PREVENTIVE MAINTENANCE(GUIDE NUMBER) 23

**EQUIPMENT:** Air Compressors

**MANUFACTURER:** Vilbiss Company/Quincy/Johnson/Champion/Penn

**CATEGORY:** Air Compressors **FREQUENCY:** Monthly

**PREVENTIVE MAINTENANCE DESCRIPTION:**

Inspect air compressor, check oil, PSI switch, and wipe down unit.

**SPECIAL INSTRUCTIONS:**
Any needed repairs should be reported for correction at a later date. When oil level is found to be low, add oil the same day of inspection.

**PROCEDURE:**

Inspect compressor unit for cleanliness:

1. Visually inspect compressor unit for cleanliness: clean the fins on the cylinders, heads, intercoolers, and aftercoolers. Wipe down compressor, motor, and tank.

2. Remove cover of pressure regulator switch and examine contract points for damage; examine interior of switch for cleanliness. Clean if needed. Replace and secure cover.

3. Drain tank if unit is not equipped with an automatic drain on the lowest point of the tank. Blow down automatic control condensate traps when unit is equipped for the function.

4. Operate safety valves to insure they relieve and reset without sticking.

5. Make sure compressor is operating at the proper pressure setting(s) and has the proper oil level.

**TOOLS AND MATERIALS:** **Revision Date:** 11/07/88
Standard hand tools, rags.

**Figure 4.13** PM Guide Number 23—Air compressors. (*From Georgetown University.*)

GEORGETOWN UNIVERSITY PREVENTIVE MAINTENANCE(GUIDE NUMBER) 28

EQUIPMENT: Air Compressor, Heavy Duty

MANUFACTURER: Pennsylvania Pump & Compressor Company

CATEGORY: Air Compressors FREQUENCY:

PREVENTIVE MAINTENANCE DESCRIPTION:

Change crank case oil(Pennsylvania Compressors).

SPECIAL INSTRUCTIONS:

Follow OSHA Standard 1910.213- Open lock and tag motor disconnect switch before starting work.

PROCEDURE:

1. Change crank case oil: (Gulf Harmony 53; 13 Quarts for air compressors)

NOTE: If the oil in the frame oil level gauge darkens and takes on a dirty appearance before scheduled replacement, it is to be changed at that time.

a. Stop the compressor and open the motor disconnect switch.

b. Remove oil drain plug from end of drain line and drain old oil into waste container.

c. Remove Air Breather unit, clean, oil, and reassemble it.

d. Thoroughly clean inside of frame sump using kerosene or a solvent before refilling with new oil.

e. After the compressor is running, check the oil level to ensure that it is at the halfway mark of the sight gauge on the bed.

NOTE: Lubricating Oils(General):
FRAME: The lubricant for the splash system should be a neutral mineral oil and inhibited against oxidation and corrosion, corresponding to the SAE 30 or 40 grade. Heavy duty oils are also satisfactory lubricants and generally will provide results somewhat superior to straight mineral oils.

Page 1 of 2

**Figure 4.14** PM Guide Number 28—Air compressors, heavy-duty. (*From Georgetown University.*)

GEORGETOWN UNIVERSITY PREVENTIVE MAINTENANCE(GUIDE NUMBER) 28

**PROCEDURE, continued-**

CYLINDER LUBRICATING OILS(GENERAL) FOR AIR COMPRESSORS ONLY:

Air compressors, due to the moisture in the air, require the use of an oil compounded with 3 to 5% of acidless tallow that contains an oxidation and rust inhibitor and has a low carbon content.

A high grade, 100% distilled , solvent refined, straight mineral oil containing an oxidation, corrosion or rust inhibitor and of an SAE 30 or 40 weight will provide satisfactory compressor cylinder lubrication. It should have good polar characteristics, good wetting ability, high film strength, good chemical stability, be resistant to sludging and should contain as low a carbon content as possible.

**TOOLS AND MATERIALS:** Revision date: 11/07/88
Standard hand tools, waste oil container, 13 quarts of oil.

**Figure 4.14** *(Continued)*

GEORGETOWN UNIVERSITY PREVENTIVE MAINTENANCE(GUIDE NUMBER) 32

EQUIPMENT: Fan Coil Unit; Ceiling-Mounted; Sizes 1 through 7

MANUFACTURER:

CATEGORY: Fan Coil Units FREQUENCY:

PREVENTIVE MAINTENANCE DESCRIPTION:

Clean fan coil unit. Re-oil fan motor.

SPECIAL INSTRUCTIONS:

Covers should be securely bolted to unit to prevent tampering by tenants(Replace missing fasteners).

PROCEDURE:

1. Clean fan coil unit:

   a. Open front access panel and remove filter.

   b. Brush(or vacuum) coils and fins.

   c. Clean the drain pan and check drain line opening for debris.

   d. Wipe the entire casing clean and inspect for rust.

   e. Replace or clean the filter.

2. Re-oil fan motor:

   a. Insert 10 to 15 drops of oil into each of the motor sleeve bearings through the oil extension lines.

   b. Wipe motor fan blades and casing clean: manually rotate fan and feel for debris.

   c. Energize motor and check for excessive vibration and noise.

   d. Check wiring for loose or unprotected connections. Test operating controls.

TOOLS AND MATERIALS: Oil can, hand tools, rags, brush or vacuum, spare fasteners.

**Figure 4.15** PM Guide Number 32—Fan coil unit, ceiling-mounted. (*From Georgetown University.*)

**GEORGETOWN UNIVERSITY PREVENTIVE MAINTENANCE(GUIDE NUMBER)** 33

**EQUIPMENT:** Fan Coil Units

**MANUFACTURER:**

**CATEGORY:** **FREQUENCY:**

**PREVENTIVE MAINTENANCE DESCRIPTION:**

Change or clean air filters, clean condensate drains and pans, clean fan and coils, oil fan (FCU's)

**SPECIAL INSTRUCTIONS:**

Turn off power to unit before starting work.

**PROCEDURE:**

A. Remove access covers from unit, remove filter. Use a soft brush to clean coil fins and the fan blade(s). Vacuum inside of unit thoroughly.

B. Clean out any scale or dirt accumulation in drip pan(s).

C. If the unit has a trap on the condensate line add 4 ounces of drain cleaner to the trap. (Do not use acid type drain cleaners-use enzyme type cleaner such as Mr. Plumber)

D. Add 6 drops of SAE 20 (ND) oil to each bearing on the fan motor.

E. Start Fan Motor make sure unit is running smoothly. If there is excessive vibration or noise report it to supervisor so it can be scheduled for repair at a later date.

F. reinstall clean filter (ensure that filter fits securely and will not fall out).

G. Replace cover on fan coil unit. Covers should be securely bolted to unit to prevent tampering by tenants. (Replace any missing fasteners.)

**TOOLS AND MATERIALS:** **Revision Date:** 10/27/88
Standard Hand Tools, Vacuum, Rags, Filters

**Figure 4.16** PM Guide Number 33—Fan coil units. (*From Georgetown University.*)

GEORGETOWN UNIVERSITY PREVENTIVE MAINTENANCE(GUIDE NUMBER) 35

EQUIPMENT: Condensers

MANUFACTURER: All

CATEGORY: Air conditioning FREQUENCY: Yearly

PREVENTIVE MAINTENANCE DESCRIPTION:

Package airconditioning unit spring check.

SPECIAL INSTRUCTIONS:

OSHA Standard 1910.213 -Open lock and tag motor disconnect switch before starting work.

PROCEDURE:

A. Remove panels. Thoroughly inspect and clean interior and exterior of machine.
B. Clean the condenser by hosing with water. An approved coil cleaner may also be used if necessary. **Do not use hot water or steam.**
C. Carefully inspect all electrical terminals for signs of looseness or overheating.
D. Lubricate motor and fan bearings.
E. Replace belts on condenser fan if so equipped.
F. Start unit and check the refrigerant level. Recharge if needed. Report leaks to supervisor.
G. Check the oil level on compressors equipped with sight glasses.
H. Tighten any fan and compressor motor mounting bolts if found to be loose.
I. Check for proper operation of relays, switches, and safety devices.
J. Check amperages of compressor and fan motors. Compare with nameplate date to see that they are within their proper values.

TOOLS AND MATERIALS: Revision Date: 11/07/88
AC Manifold set, freon, oil-can with SAE 30 ND oil, grease gun, garden hose, hand tools to remove covers.

**Figure 4.17** PM Guide Number 35—Condensers. (*From Georgetown University.*)

**GEORGETOWN UNIVERSITY PREVENTIVE MAINTENANCE(GUIDE NUMBER)** 46

**EQUIPMENT:** AIR HANDLER UNIT: SUPPLY (DEHUMIDIFIER)

**MANUFACTURER:**

**CATEGORY:** **FREQUENCY:**

**PREVENTIVE MAINTENANCE DESCRIPTION:**

1. Clean and adjust spray type dehumidifier unit.

**SPECIAL INSTRUCTIONS:**

OSHA Standard 1910.213 -Open lock and tag motor disconnect switch before starting work.

**PROCEDURE:**

1. Clean and adjust spray type dehumidifier unit.

   a. Drain sump tank and scrub clean of any algae, chemical residue, or sediment. Remove and clean spray pump return-line strainer and check for condition.

   b. Flush out tank thoroughly; close drain valves and fill tank.

   c. Check operation and condition of float valve. Observe water level when float valve closes; it should be ½" below the tank overflow outlet. Adjust float, if necessary.

   d. Check spray nozzles for proper functioning; remove and clean, if required.

   e. Check spray pressure gauge for correct pressure of 10 psi. These units are to be operated continuously.

   263-SB09-023 (AHU-8)
   051-P1-023 (AHU-3)
   051-P1-022 (AHU-4)
   051-P1-024 (AHU-5)

**TOOLS AND MATERIALS:** **Revision Date:** 11/07/88
Hose, bucket, hand tools, wire brush, 18" and 24" pipe wrenches.

**Figure 4.18** PM Guide Number 46—Air handler unit, supply (dehumidified). (*From Georgetown University.*)

GEORGETOWN UNIVERSITY PREVENTIVE MAINTENANCE(GUIDE NUMBER) 47

EQUIPMENT: AHU's

MANUFACTURER: All

CATEGORY: Cleaning FREQUENCY: As Needed

PREVENTIVE MAINTENANCE DESCRIPTION:

Replace throwaway filters

SPECIAL INSTRUCTIONS:

OSHA Standard 1910.213 -Open lock and tag motor disconnect switch before starting work.

PROCEDURE:

A. Turn off air handling unit and remove old filters.

B. Install new filters. Insure that there are no gaps between media.

C. Reinstall any spacers and restart unit.

Note: Two indicators used to determine the need for servicing are a 10% decrease in air flow or an increase in resistance of two to three times the initial resistance. See the table below for some typical velocities and air resistances of various filters.

Clean Filter Operating Data

| Type of air cleaner | Nominal velocity through media(fpm) | Resistance through clean filter(in. wg) |
|---|---|---|
| Viscous Impingement | | |
| throwaway (2 in.) | 300 | 0.06 - 0.12 |
| renewable (4 in.) | 300 | 0.12 - 0.24 |
| cleanable (2 in.) | 300-500 | 0.04 - 0.12 |
| (4 in.) | 300 | 0.08 - 0.20 |
| automatic self-cleaning | 500 | 0.30 - 0.50 |

Pg 1 of 2

**Figure 4.19** PM Guide Number 47—Air handler units (all). (*From Georgetown University.*)

GEORGETOWN UNIVERSITY PREVENTIVE MAINTENANCE(GUIDE NUMBER) 47

**Clean Filter Operating Data** (cont.)

| Type of air cleaner | Nominal velocity through media(fpm) | Resistance through clean filter(in. wg) |
|---|---|---|
| Dry media | | |
| -cleanable and | | |
| renewable (2 in.) | 60 | 0.08 - 0.13 |
| (8 in.) | 35 | 0.10 - 0.12 |
| high efficiency renewable | 5-20 | 0.50 - 1.20 |
| Electronic(ionizing) | | |
| plate or cell | 300-400 | 0.15 - 0.30 |
| automatic | 400-500 | 0.20 - 0.32 |
| Electronic charged media | 35 | 0.03 - 0.12 |

**TOOLS AND MATERIALS:** Revision Date: 8/11/88
Drop light, replacement filters, trash bags, screw driver, channel locks.

Pg 2 of 2

**Figure 4.19** (*Continued*)

GEORGETOWN UNIVERSITY PREVENTIVE MAINTENANCE(GUIDE NUMBER) 51

EQUIPMENT: Pump Coupling

MANUFACTURER:

CATEGORY: FREQUENCY:

PREVENTIVE MAINTENANCE DESCRIPTION:

1. Check flexible coupling for pump/motor alignment.

2. Check flexible coupling for condition.

SPECIAL INSTRUCTIONS:

PROCEDURE:

1. Check flexible coupling for pump/motor alignment:

   a. Open motor disconnect switch or starter / controller contacts.

   b. To check angular alignment, measure the distance between the coupling flanges at 90° intervals. If this spacing is unequal, this is angular misalignment.

   c. To check for parallel alignment, lay a 6" steel scale across the outside flanges of the coupler halves. If the scale does not lay flat on both flange edges, parallel misalignment exists.

2. Check flexible coupling for condition:

   a. Visually check flexible insert for damage or wear. If necessary, loosen coupling set-screws and slide coupler halves back on shaft; remove flexible part of coupling and inspect it for damage and excessive wear.

   b. Reinstall the flexible part then check coupling hubs for loose shaft keys and set-screws.

TOOLS AND MATERIALS: Revision Date: 11/09/88

**Figure 4.20** PM Guide Number 51—Pump coupling. (*From Georgetown University.*)

GEORGETOWN UNIVERSITY PREVENTIVE MAINTENANCE(GUIDE NUMBER) 52

EQUIPMENT: Centrifugal Pump (not integral with motor)

MANUFACTURER:

CATEGORY: FREQUENCY:

PREVENTIVE MAINTENANCE DESCRIPTION:

1. Check operating pump.

2. Check NON-operating pump.

SPECIAL INSTRUCTIONS:

PROCEDURE:

1. Check operating pump:
   a. While pump is in operation, note performance, bearing temperature, pressure gauge and operation of any installed stuffing boxes. (compare readings against previous readings to determine need for gauge calibration or repair work.) Conduct static pressure check, when possible.

   b. Check mechanical seal or gland seal leak-off. If gland leak-off is excessive and cannot be reduced to a stream about the size of a pencil lead, it is necessary that the box be repacked.

   c. Stop and start pump, noting undue vibration, noise, pressure, and the action of the check valve.

   d. Upon completion of test, adjust any gland leak-off to a stream about the size of a lead pencil.

   e. Check the overflow drain hole for foreign matter; clean it out as necessary.

   f. Inspect unit for corrosion or damaged paint work.

Note**-When stuffing box repacking is necessary, close the hand suction and discharge valves, drain pump casings and note if valves are holding properly.

Pg 1 of 2

**Figure 4.21** PM Guide Number 52—Centrifugal pump (not integral with motor). (*From Georgetown University.*)

**GEORGETOWN UNIVERSITY PREVENTIVE MAINTENANCE(GUIDE NUMBER) 52**
-continued.

2. Check NON-operating pump:

   a. Remove packing glands. Clean glands and adjusting studs with wire brush; coat both with grease.

   b. Check stuffing box packing for damage; replace if necessary.

   c. Clean pump strainer.

   d. Check the overflow drain hole for foreign matter; clean it out as necessary.

   e. Inspect unit for corrosion or damaged paint work.

---

**TOOLS AND MATERIALS:** Revision Date: 11/09/88

**Figure 4.21** *(Continued)*

GEORGETOWN UNIVERSITY PREVENTIVE MAINTENANCE(GUIDE NUMBER) 55

EQUIPMENT: Sump Pump

MANUFACTURER:

CATEGORY: FREQUENCY:

PREVENTIVE MAINTENANCE DESCRIPTION:

1. Relubricate intermediate and casing sleeve bearings.

2. Relubricate float-rod felt washer.

SPECIAL INSTRUCTIONS:

PROCEDURE:

1. Relubricate intermediate and casing sleeve bearings:

   a. Clean off the zerk fittings on the floor plate and attach hand grease gun.

   b. Insert Gulfcrown Special grease until a "back pressure" is felt, which indicates a filled bearing cavity.

2. Relubricate float-rod felt washer:

   a. Saturate the felt washer by applying Gulf Harmony 69 oil to the float rod and allowing it to drain into the guide pipe in the basin cover.

TOOLS AND MATERIALS: Revision Date: 10/20/88

**Figure 4.22** PM Guide Number 55—Sump pump. (*From Georgetown University.*)

GEORGETOWN UNIVERSITY PREVENTIVE MAINTENANCE(GUIDE NUMBER) 56

EQUIPMENT: Sump Pump

MANUFACTURER:

CATEGORY: FREQUENCY:

PREVENTIVE MAINTENANCE DESCRIPTION:

1. Relubricate thrust ball bearing.
2. Check flexible coupling.
3. Test high water alarm.

SPECIAL INSTRUCTIONS:

PROCEDURE:

1. Relubricate Pump Thrust Bearings.

a. Wipe lubrication fitting clean; inject grease until a back pressure can be felt. Wipe up any excess grease.

2. Check Flexible Coupling.

a. Inspect flexible coupling for wear, damage and alignment.

3. Test High Water Alarm.(test all sumps in the same space simultaneously)

a. Open pump motor disconnect switch and allow pump basin to fill with liquid.

b. When high water alarm sounds, close disconnect switch and observe pump and float switch for normal operation.

c. When no high water alarm exists, lift the float switch arm manually and observe switch for normal operation. Note: Do not drop the float switch arm but lower it gently until motor is de-energized.

d. Observe the motor to see if it comes quickly up to speed and maintains a constant rotation rate.

e. Be alert for unusual or excessive noise. Note that Weil F2 type pumps have a coarse strainer, be alert for symptoms of partial clogging.

TOOLS AND MATERIALS: Revision Date: 11/02/88

**Figure 4.23** PM Guide Number 56—Sump pump. (*From Georgetown University.*)

EQUIPMENT: Condensate Pump: type HS, Horizontal Centrifugal,V-S

MANUFACTURER:

CATEGORY: FREQUENCY:

PREVENTIVE MAINTENANCE DESCRIPTION:

1. Regrease pump ball bearings.
2. Check float switch and control assembly.

SPECIAL INSTRUCTIONS:

PROCEDURE:

1. Re-grease pump ball bearings.

a. With the pump stopped, clean off the bearing housing and seal.
b. Start pump and inject Gulf Crown Grease EP Special, or Equal, into the lubrication fitting until a slight bead of grease appears at the bearing seal. Caution: do not overgrease.
c. Check metallic packing for proper leak-off.

2. Check float switch and control assembly:

a. Observe that the float switch opens and closes properly as receiver fills and is emptied by the pump. Check float control linkage for binding.

TOOLS AND MATERIALS: Revision Date: 10/20/88

**Figure 4.24** PM Guide Number 58—Condensate pump, type HS, horizontal centrifugal, V-S. (*From Georgetown University.*)

**GEORGETOWN UNIVERSITY PREVENTIVE MAINTENANCE(GUIDE NUMBER)** 61

**EQUIPMENT:** Condensate pump

**MANUFACTURER:**

**CATEGORY:** **FREQUENCY:** Yearly

**PREVENTIVE MAINTENANCE DESCRIPTION:**

1. Drain and flush condensate receiver. Inspect internal parts of check valve test pump and float switches.

**SPECIAL INSTRUCTIONS:**
OSHA Standard 1910.213 -Open lock and tag motor disconnect switch and starter before starting work. Check the voltage of all incoming line terminals. Positively ascertain that the equipment is deenergized.

**PROCEDURE:**

1. a. Close valves in discharge and return lines; open drain and empty receiver, then flush receiver to remove sediment and scale.
   b. Check temperature of condensate (it should be approximately 30° below steam temperature if traps are not leaking).
   c. Open check valve and inspect valve disk and seat for cuts, pitting, sediment of scale deposits. Inspect valve stem and pin for wear. As an alternate: listen to valve operate.
   d. Turn shaft of pumps and see that they rotate freely by hand.
   e. Remove float switch cover and examine interior for damage, moisture, oil, and dirt. Push down on float switch arm to check that it operates pump.
   f. When starting up, open valves in discharge and return lines, close drain.
   g. For Reiss Science: Remove strainer clean-out cover from side of receiver; remove strainer and clean it.
   h. Check motor bearings for unusual noise.

**TOOLS AND MATERIALS:** **Revision Date:** 11/02/88

**Figure 4.25** PM Guide Number 61—Condensate pump. (*From Georgetown University.*)

**GEORGETOWN UNIVERSITY PREVENTIVE MAINTENANCE(GUIDE NUMBER)** 63

**EQUIPMENT:** Pump Ball Bearings

**MANUFACTURER:**

**CATEGORY:** **FREQUENCY:** Yearly

**PREVENTIVE MAINTENANCE DESCRIPTION:**

1. Add grease to pump bearings.

**SPECIAL INSTRUCTIONS:**

Excess grease is the most common cause of overheating and bearing damage. Avoid adding too much grease.

**PROCEDURE:**

1. Add grease to pump bearings.

a. Clean off the lubricating fittings and inject approximately 1/2 ounce of grease (about a teaspoonful for bearings of small size, and a tablespoonful for larger sizes). The bearing housing is intended to only be 1/4 to 1/3 full of grease.

b. Wipe all visible grease from fittings and bearing housing.

**TOOLS AND MATERIALS:** **Revision Date:** 11/02/88
Grease gun, No 2 grease, Rags, Hand tools.

**Figure 4.26** PM Guide Number 63—Pump ball bearings. (*From Georgetown University.*)

GEORGETOWN UNIVERSITY PREVENTIVE MAINTENANCE(GUIDE NUMBER) 66

EQUIPMENT: Radiators, Heating

MANUFACTURER:

CATEGORY: FREQUENCY:

PREVENTIVE MAINTENANCE DESCRIPTION:

1. Clean and check hot water radiator (convector and induction type).

SPECIAL INSTRUCTIONS:

PROCEDURE:

1. Convector type:
   a. Remove front panel and vacuum or brush dirt out of unit and clear of fins.
   b. Check bleed valve for condition.
   c. Check radiator valve for free turning and seating. Check packing.
   d. Wipe the entire unit clean and check for rust spots.
2. Induction type:
   a. Vacuum clean the coil fins. If necessary, remove cover to get at coils.
   b. Check for loose fittings or damage.
   c. Wipe the entire unit clean and inspect for rust spots.

TOOLS AND MATERIALS: Revision Date: 10/21/88

**Figure 4.27** PM Guide Number 66—Radiators, heating. (*From Georgetown University.*)

**GEORGETOWN UNIVERSITY PREVENTIVE MAINTENANCE**(GUIDE NUMBER) 69

**EQUIPMENT:** Valves (excluding sprinkler systems)

**MANUFACTURER:** All

**CATEGORY:** Lubrication **FREQUENCY:** Yearly

**PREVENTIVE MAINTENANCE DESCRIPTION:**

Inspect and service valve packing and lube stem.

**SPECIAL INSTRUCTIONS:**

After servicing leave the valve in the position it was found in.

**PROCEDURE:**

A. Exercise valve from one limit to the other (open and closed). Lightly lubricate stem with graphite.

B. Adjust packing gland to stop any leakage.

C. Any leaks that can not be repaired during this PM whether at the packing or elsewhere is to be noted on the work order and also reported to the shop foreman.

**TOOLS AND MATERIALS:** **Revision Date:** 11/09/88
Graphite, Flashlight, 9 ft. ladder.

**Figure 4.28** PM Guide Number 69—Valves (excluding sprinkler systems). (*From Georgetown University.*)

GEORGETOWN UNIVERSITY PREVENTIVE MAINTENANCE(GUIDE NUMBER) 72

EQUIPMENT: Belt Drive Alignment

MANUFACTURER: Trane

CATEGORY: Power Transmission FREQUENCY: Yearly

PREVENTIVE MAINTENANCE DESCRIPTION:

Belt Drive Alignment Check

SPECIAL INSTRUCTIONS:

OSHA Standard 1910.213 -Open lock and tag motor disconnect switch before starting work.

PROCEDURE:

A. Sheave Alignment:

Alignment of the belt sheaves is crucial to V-Belt life and should be checked periodically. Align the fan and motor sheaves by using a straightedge that is long enough to span the distance between the outside edges of the sheaves. When sheaves are aligned the straightedge will touch both sheaves squarely across their face. For uneven width sheaves a string drawn tight through the center groove of both sheaves with an equal measurement made from each end of the string to an outboard parallel point will also give a good indication of proper alignment.

When alignment needs to be changed during this PM care should be taken to ensure that set screws are retorqued to their recommended values.

Torques for Tightening Set Screws

| Set Screw Diameter | Hex Size Across Flats | Recommended Torque Inch Pounds | Recommended Torque Foot Pounds |
|---|---|---|---|
| 1/4 | 1/8 | 66 | 5.5 |
| 5/16 | 5/32 | 126 | 10.5 |
| 3/8 | 3/16 | 228 | 19.0 |
| 7/16 | 7/32 | 348 | 29.0 |
| 1/2 | 1/4 | 504 | 42.0 |
| 5/8 | 5/16 | 1104 | 92.0 |

OOLS AND MATERIALS:
traightedge, String, 12 Foot Ruler, Allen Hex Set, Torque Wrench.

**Figure 4.29** PM Guide Number 72—Belt drive alignment. (*From Georgetown University.*)

**GEORGETOWN UNIVERSITY PREVENTIVE MAINTENANCE(GUIDE NUMBER)** 74

**EQUIPMENT:** Dampers

**MANUFACTURER:** All

**CATEGORY:** Lubrication **FREQUENCY:** Yearly

**PREVENTIVE MAINTENANCE DESCRIPTION:**

Dampers, Linkage, and Actuators care and maintenance

**SPECIAL INSTRUCTIONS:**

OSHA Standard 1910.213 -Open lock and tag motor disconnect switch before starting work.

**PROCEDURE:**

A. Dampers and their associated drive components should be inspected periodically for freedom of movement. Also check bolts, clip screws, and locknuts for tightness. Dampers with plastic sleeve type bearings do not require lubrication and should only be cleaned with a dry cloth as conditions dictate. Other types of bushings may be lubricated lightly. Remove any excess lubricant to prevent dirt collecting and binding damper.

B. Check outside air inlet screens during this PM. Clean any debris from screen.

**TOOLS AND MATERIALS:** Rags for cleaning, Oil Can (SAE 20 [ND] Oil), Hand Tools.

**Figure 4.30** PM Guide Number 74—Dampers. (*From Georgetown University.*)

GEORGETOWN UNIVERSITY PREVENTIVE MAINTENANCE(GUIDE NUMBER) 79

EQUIPMENT: Hot Water Heater (heat exchanger)

MANUFACTURER:

CATEGORY: FREQUENCY:

PREVENTIVE MAINTENANCE DESCRIPTION:

1. Check the temperature of the exit water of the heater.

SPECIAL INSTRUCTIONS:

PROCEDURE:

1. Check the temperature of the exit water of heater:

a. If temperature of exit water is 120°F, or less, then schedule the below descaling procedure for accomplishment: the rapid descaling method consists of the following procedure.

1. Close valves in the inlet and outlet water lines. Also drain the shell through the blowdown valve while opening the shell to atmospheric pressure through the relief valve connection.
2. Close High Temperature return line by means of hand valve.
3. Inject a small flow of cold water through the open relief valve connection, allowing the drain connection to remain open at the bottom of the heater.
4. Shut off cold water flow through the open relief valve connection and open high temperature water flow into coils. After two minutes interval, close high temperature water flow into coils and open cold water flow through open relief valve connection. Repeat this procedure for several minutes in approximately two minute intervals.
5. Open the main line of water supply into the heater and allow a complete flushing action of the solid particles dislodged from the coils to be discharged through the blowdown connection.
6. Restore the heater into service by opening those valves which had been closed and closing the relief valve and blowdown connection.

**Figure 4.31** PM Guide Number 79—Hot water heater (heat exchangers). (*From Georgetown University.*)

GEORGETOWN UNIVERSITY PREVENTIVE MAINTENANCE(GUIDE NUMBER) 79
-continued

This method of descaling consists essentially of thermal shocking the coiled heating surfaces by alternately heating and cooling. This causes them to expand and contract freely because of the design of the coil structure and the method of suspension. No damage is done to the internal parts of the heater during this procedure because the heater is designed to permit this continual sudden expansion and contraction.

It has been proven by experience, that scale formation is very rapid when:

1. The quantity of water is large (being passed through the heater).

2. The temperature of the water is elevated beyond 150°F;

3. High pressures are used in the coils (which is the severe condition in hard water areas); a treatment of a few minutes duration as explained above, has successfully restored heaters to over 90% of their design capacities.

TOOLS AND MATERIALS: Revision Date: 10/21/88

Pg 2 of 2

Figure 4.31 *(Continued)*

GEORGETOWN UNIVERSITY PREVENTIVE MAINTENANCE(GUIDE NUMBER) 80

EQUIPMENT: Steam Converters

MANUFACTURER:

CATEGORY: FREQUENCY:

PREVENTIVE MAINTENANCE DESCRIPTION:
1. Inspect converter for leakage.
2. Clean steam strainer.
3. Check pressure relief valve.

SPECIAL INSTRUCTIONS:

PROCEDURE:

1. Inspect converter for leakage.

a. Inspect external areas and piping connections for leakage.

b. Check for evidence of internal leakage between tubes and shell: Water leaking into steam system causes continuous flow of water through the steam trap.

c. When exit water temperature drops to about 150°F, schedule converter for chemical cleaning of the water sides only.

d. Inspect gauges and thermometers for leaks, broken glass, damaged thermometer bodies, inaccurate readings or faulty operation due to defective parts.

2. Clean Steam strainer:

a. remove basket from strainer and clean.

b. Check operation of relief valve by lifting lever and releasing it. Observe valve for positive seating.

TOOLS AND MATERIALS: Revision Date: 10/21/88

**Figure 4.32** PM Guide Number 80—Steam converters. (*From Georgetown University.*)

GEORGETOWN UNIVERSITY PREVENTIVE MAINTENANCE(GUIDE NUMBER) 81

EQUIPMENT: Domestic Hot Water Heater

MANUFACTURER:

CATEGORY: FREQUENCY:

PREVENTIVE MAINTENANCE DESCRIPTION:

1. Flush and inspect Hot Water Heater Tank.
2. Clean Hot Water Tank.

SPECIAL INSTRUCTIONS:

PROCEDURE:

1. Flush and inspect Hot Water Heater Tank:

a. Actuate lever of pressure relief valve to test for operation.

b. Close steam and water valves, release pressure on water side of unit and drain water out of vessel.

c. Remove manhole cover, flush tank out and inspect the heating surface for fouling and buildup of scale or mineral deposits; mechanically clean heating surfaces or schedule chemical cleaning, if required.

d. Inspect internal shell for resting or pitting.

e. Remove anode (if installed) and clean by flaking off decomposed layer with light metal object. Schedule anode for replacement if more than 50% has decomposed.

f. Clean steam strainer.

g. Replace manhole gasket; reinstall manhole cover, refill tank and check for leaks.

2. Shock Procedure:

-Subject unit to a shock treatment as prescribed by the foreman.

TOOLS AND MATERIALS: Revision Date: 10/21/88

**Figure 4.33** PM Guide Number 81—Domestic hot water heater. (*From Georgetown University.*)

GEORGETOWN UNIVERSITY PREVENTIVE MAINTENANCE(GUIDE NUMBER) 82

EQUIPMENT: Steam Safety Relief Valves

MANUFACTURER: All

CATEGORY: Life Safety FREQUENCY: Semi-annual

PREVENTIVE MAINTENANCE DESCRIPTION:

Test and inspect steam pressure relief valve.

SPECIAL INSTRUCTIONS:

PROCEDURE:

Lift the operating lever on the safety valve. Check for the following:

A. That the valve opens and resets properly.

B. The valve closes tightly without chattering.

C. That the valve is tightly closed and there is no leakage of steam.

D. Examine the valve visually. See that there has been no tampering.

E. Check the atmospheric discharge piping, see that there is proper drainage.

Note- Atmospheric pipe drainage prevents a column of water from building up over the safety relief valve disk and thus increasing the popping pressure. Water can also produce an explosive-like slug of water being fired out of the valve.

TOOLS AND MATERIALS: Revision Date: 8/15/88
Flashlight, ladder( for certain locations).

**Figure 4.34** PM Guide Number 82—Steam safety relief valves. (*From Georgetown University.*)

GEORGETOWN UNIVERSITY PREVENTIVE MAINTENANCE(GUIDE NUMBER) 84

EQUIPMENT:Air Compressor,Heavy Duty,Horizontal,Rcprctng,9"stroke

MANUFACTURER:

CATEGORY: FREQUENCY:

PREVENTIVE MAINTENANCE DESCRIPTION:

1. Check rider rings and piston rings for wear.

SPECIAL INSTRUCTIONS:

PROCEDURE:

1. Check rider rings and piston rings for wear:

a. In conjunction with PM Guide, remove back head from air cylinder.

b. Inspect interior of cylinder for grit and rust. All impurities must be thoroughly removed from cylinder to prevent scoring of the cylinder walls.

c. Rider ring check: Use a feeler gauge to check clearance between piston and bore. The gap at the bottom of the cylinder should not be less than 0.02 inch.

d. Piston ring check; loosen crosshead nut and unscrew piston until piston ring gap is visible in the bore when viewed through a valve hole. Use a feeler gauge to determine amount of ring gap. The minimum permissible gap between the ends of the ring is 0.325 inch.

e. Remove the two bolts from the packing gland and move connecting rod so packing moves into view. Visually check packing ring gap, if it is almost closed it is to be replaced.

TOOLS AND MATERIALS: Revision Date: 10/21/88

**Figure 4.35** PM Guide Number 84—Air compressor, heavy-duty, horizontal, reciprocating, 9-in stroke. (*From Georgetown University.*)

GEORGETOWN UNIVERSITY PREVENTIVE MAINTENANCE(GUIDE NUMBER) 89

EQUIPMENT: Cooling tower

MANUFACTURER:

CATEGORY: FREQUENCY:

PREVENTIVE MAINTENANCE DESCRIPTION:

1. Add lubricating oil to fan bearing.
2. Check operating components.

SPECIAL INSTRUCTIONS:

PROCEDURE:

1. Add lubricating oil to fan bearing:

   a. Basic Science; Stop fan motor; fill oil cups of fan bearings with Gulf Harmony 97 oil, or equal.

2. Check operating components:

   a. Check for correct water level in basin while pump is operating.

   b. Check pump suction strainer-screen for cleanliness.

   c. Check spray nozzle operation (if applicable).

   d. Check belt condition.

TOOLS AND MATERIALS: Revision Date: 10/27/88

**Figure 4.36** PM Guide Number 89—Cooling tower. (*From Georgetown University.*)

**GEORGETOWN UNIVERSITY PREVENTIVE MAINTENANCE(GUIDE NUMBER) 93**

**EQUIPMENT:** Fire Extinguishers

**MANUFACTURER:**

**CATEGORY:** **FREQUENCY:**

**PREVENTIVE MAINTENANCE DESCRIPTION:**

1. Conduct monthly fire extinguisher inspection.

**SPECIAL INSTRUCTIONS:**

**PROCEDURE:**

1. Conduct monthly fire extinguisher inspection:
   a. Use the inventory list of fire extinguisher locations in the designated building and conduct this inspection on each extinguisher located at the stations.
   b. Check each extinguisher location to ensure that the correct type of fire extinguisher is at the station. (Extinguisher type is indicated on the Location List).
   c. Inspect extinguisher to ensure that it has not been used or tampered with: water and air- check air pressure; soda-acid check nozzle for chemical residue; Co2-check wire seal to ensure that it is not broken.
   d. Inspect for any obvious physical damage, corrosion or hose impairment.
   e. Check cabinets for accessibility, operable handle, broken glass, and trash.

Note- The tag attached to the extinguisher is not to be initialed for a monthly "inspection". This tag is for recording the yearly "maintenance action" or a "recharge date" and is to be initialed at those times.

NFPA No 10A -1970 Paragraph 1200 is quoted to further define an "Inspection":

> "An inspection is a 'quick check' that an extinguisher is available and will operate. It is intended to give reasonable assurance that the extinguisher is fully charged and operable. This is done by seeing that it is in its designated place, that it has not been actuated or tampered with, and that there is no obvious physical damage or condition to prevent operation."

-Paragraph 1230: "The whole intent of an inspection is to find out quickly if something is wrong so that proper corrective action can be taken."

**TOOLS AND MATERIALS:** Revision Date:10/28/88

**Figure 4.37** PM Guide Number 93—Fire extinguishers. (*From Georgetown University.*)

GEORGETOWN UNIVERSITY PREVENTIVE MAINTENANCE(GUIDE NUMBER) 94

EQUIPMENT: Fire Extinguisher

MANUFACTURER:

CATEGORY: FREQUENCY:

PREVENTIVE MAINTENANCE DESCRIPTION:

1. Conduct annual fire extinguisher maintenance.

2. Verify fire extinguisher serial number and hydrostatic test equipment.

SPECIAL INSTRUCTIONS:

PROCEDURE:

1. Conduct monthly fire extinguisher maintenance:

a. Check for physical damage, dents, corrosion, defective nozzle/horn or hose.
b. Check to ensure that the lettering on instructions is easy to read.
c. Check extinguisher hanger for firm mounting, or that extinguisher cabinet door is operable.
d. Sign the attached inspection tag upon completion of maintenance.

-FOR WATER AND AIR EXTINGUISHERS:

a. Check gauge reading to see that it is in proper range.

-FOR Co2 EXTINGUISHERS:

a. Check wire seal to ensure that it is not broken.
b. Weigh extinguisher and compare to amount stamped on bottle.

-FOR SODA AND ACID EXTINGUISHER:

a. Remove, disassemble, and recharge according to manufacturer's instructions.

Caution- Do not invert (pressurize) any extinguisher that shows signs of mechanical damage or corrosion.

Pg 1 of 2

**Figure 4.38** PM Guide Number 94—Fire extinguishers. (*From Georgetown University.*)

**GEORGETOWN UNIVERSITY PREVENTIVE MAINTENANCE**(GUIDE NUMBER) 94
-continued

2. Verify fire extinguisher serial number and hydrostatic test equipment:

a. Verify the serial number on the PM Card against the extinguisher serial number; if they are different,fill out a "Change Form" for computer updating of correct number.

b. Check the last hydrostatic test date; if it is more than four years ago, schedule the extinguisher for a retest to coincide with the 5th year date.

c. Do not hydrostatic test Soda Acid extinguishers. At time of five year testing cycle, replace the extinguishers with stored pressure type.

**TOOLS AND MATERIALS:** Revision Date: 10/27/88

**Figure 4.38** (*Continued*)

GEORGETOWN UNIVERSITY PREVENTIVE MAINTENANCE(GUIDE NUMBER) 96

EQUIPMENT: Sprinkler Valves Wet System

MANUFACTURER: OS+Y VALVES, MILWAUKEE

CATEGORY: Life Safety FREQUENCY:Monthly

PREVENTIVE MAINTENANCE DESCRIPTION:

Inspection of Fire Suppression System (valves and water pressure)monthly.

SPECIAL INSTRUCTIONS:
In the event that the system valve is found unlocked in the open or closed position, report the condition immediately to the plumbing shop foreman(after work hours report it to the Security Department). Note location of the valve and the name of the person to whom the report was given on the PM work order.

PROCEDURE:

A. Check each valve listed on the PM work order. Determine if the valve is fully open and locked.

B. If in doubt about the valve's condition, physically manipulate the valve to be sure it is open.

C. Check the gauges on the wet-pipe sprinkler system to ensure that normal water supply pressure is being maintained. A pressure reading on the gauge on the system side of an alarm valve in excess of the pressure recorded on the gauge on the supply side of the valve is normal, as the highest pressure from the supply side will get trapped in the system.
   On systems without booster pumps, an equal gauge reading could indicate a leak. If there are no visible leaks, it is possible there has been recent test run on the sprinkler system.

TOOLS AND MATERIALS: Revision Date 8/4/88
Flashlight.

**Figure 4.39** PM Guide Number 96—Sprinkler valves—wet system. (*From Georgetown University.*)

GEORGETOWN UNIVERSITY PREVENTIVE MAINTENANCE(GUIDE NUMBER) 97

EQUIPMENT: Sprinkler Valves Dry System

MANUFACTURER: OS+Y VALVES, MILWAUKEE

CATEGORY: Life Safety FREQUENCY:Monthly

PREVENTIVE MAINTENANCE DESCRIPTION:

Inspection of Fire Suppression System(dry pipe valves, air pressure, freeze protection) monthly.

SPECIAL INSTRUCTIONS:
In the event of that the system is found unlocked in the open or closed position, or the pressure gauges indicate the system is flooded with water, report the condition immediately to the plumbing shop foreman(after work hours report it to the Security Department). Note location of the valve and the name of the person to whom the report was given on the PM work order.

PROCEDURE:

A. Be sure that all system/control valves are in the open position and are locked.

B. If in doubt about the valve's condition, physically manipulate the valve to be sure it is open.

C. Check air and water pressure gauges to be certain that the required air pressure is being applied to the system.

D. During the winter months, check the sprinkler valve room heater, during each valve inspection to make sure it is operating and providing sufficient heat to prevent freezing.

Air Pressure Chart for Clapper Valves with a Six to One Operating Differential

| Water Pressure Maximum | Air Pressure Minimum | Air Pressure Maximum |
|---|---|---|
| 50 | 15 | 25 |
| 75 | 25 | 35 |
| 100 | 35 | 45 |
| 125 | 40 | 50 |

TOOLS AND MATERIALS: Revision Date: 8/4/88
Flashlight.

**Figure 4.40** PM Guide Number 97—Sprinkler valves—dry system. (*From Georgetown University.*)

GEORGETOWN UNIVERSITY PREVENTIVE MAINTENANCE(GUIDE NUMBER) 100

EQUIPMENT: Panelboards rated at 600 Volts or less

MANUFACTURER:

CATEGORY: Electrical FREQUENCY: Yearly

PREVENTIVE MAINTENANCE DESCRIPTION:

Panelboard cabinets(Boxes) annual PM.

SPECIAL INSTRUCTIONS:
Before performing any of the following operations turn off all power supplying the panelboard and check the voltage of all incoming line terminals. Positively ascertain that the equipment is deenergized.

PROCEDURE:

A. Check the panelboard for any signs of excessive operating temperature.-Enclosures not to exceed 176F, conducting parts 203F.

B. Carefully inspect all visible electrical joints and terminals. If there is any signs of looseness or overheating tighten the connections to their recommended standard torque values.

C. Clean out the panelboard. Use a brush, vacuum cleaner, or a lint-free rag. Do not use compressed air.

D. Exercise switch operating mechanisms and circuit breakers to insure their free movement.

E. Devices found in need of repair should be reported to your supervisor unless, in your opinion, any delay in having the repairs done would present a hazard to life or property. (Repair the unit)

NOTE: A comprehensive outline of NEMA approved maintenance standards is provided in this PM guide book and should be reviewed by the electrician before starting the Preventive Maintenance work order

OOLS AND MATERIALS: Revision Date:7/12/88
leaning equipment, Hand tools to remove covers, Meter to test emperatures.

**Figure 4.41** PM Guide Number 100—Panelboards rated at 600 V or less. (*From Georgetown University.*)

GEORGETOWN UNIVERSITY PREVENTIVE MAINTENANCE(GUIDE NUMBER) 108

EQUIPMENT: Sprinkler Systems

MANUFACTURER: All

CATEGORY: Safety FREQUENCY: Quarterly

PREVENTIVE MAINTENANCE DESCRIPTION:

Sprinkler system main drain flow test.

SPECIAL INSTRUCTIONS:

In the event that the test should prove to be unsatisfactory, report the condition immediately to the plumbing shop foreman. Note location of the test on the PM work order.

PROCEDURE:

A. Open the 2-inch(50-mm) drain valve, watch the supply water pressure gauge. The pressure reading will drop and stabilize. Now close the valve. The pressure reading should return to its former pressure reading quickly.

B. If the pressure reading returns slowly or not at all, this indicates there is a valve closed or some other type of obstruction is present.

C. Observe the discharged water. Watch for any heavy discoloration of the water, or obstructive material flowing from the water outlet. Discoloration or foreign material show a need for investigation to ensure there is a clear unobstructed water supply in an emergency.

TOOLS AND MATERIALS: Revision Date: 11/10/88
Flashlight, channel locks.

**Figure 4.42** PM Guide Number 108—Sprinkler system. (*From Georgetown University.*)

GEORGETOWN UNIVERSITY PREVENTIVE MAINTENANCE(GUIDE NUMBER) 109

EQUIPMENT: Fire Pumps

MANUFACTURER: Firetrol

CATEGORY: Safety FREQUENCY:

PREVENTIVE MAINTENANCE DESCRIPTION:

Fire pump inspection.

SPECIAL INSTRUCTIONS:

Note all deficiencies on PM work order.

PROCEDURE:

A. Check city water pressure and compare it to the system pressure. The system pressure should read 20 PSI higher than the city pressure.

B. Check the pump controllers. Ensure that all switches are in the energized position and set for automatic operation.

C. Report any indicator lights that are not operating to Low Voltage Shop foreman for replacement.

D. Perform (B) and (C) on the booster pump that serves the fire system.

TOOLS AND MATERIALS: Revision Date: 6/29/88

Flashlight, screwdriver.

**Figure 4.43** PM Guide Number 109—Fire pumps. (*From Georgetown University.*)

GEORGETOWN UNIVERSITY PREVENTIVE MAINTENANCE(GUIDE NUMBER) 111

EQUIPMENT: Air Compressor (centrifuge) Low Pressure

MANUFACTURER:

CATEGORY: FREQUENCY:

PREVENTIVE MAINTENANCE DESCRIPTION:

1. Conduct annual maintenance check on LP centrifugal air compressor.

SPECIAL INSTRUCTIONS:

PROCEDURE:

1. Lift pressure relief valve of air receiver tank and hold open. Observe pressure gauge to see of pump cuts in at the prescribed pressure and then pumps up tank under usage of air. Check cut-out pressure; release valve and check for proper seating. Check relief valve of separator by hand for proper operation.

   Reiss Science: Lead cut-in at 9 psi; Lead cut-out 18 psi
   Lag cut-in 8 psi; Lag cut-out 18 psi
   White Gravenor:

Check with operating personnel to ensure that pumps are alternated Lead-Lag monthly.

2. If pump capacity drops off during the above pump-up test, check for proper amount of water seal and that the compressor suction is unobstructed. (Seal water should be supplied in sufficient quantity so that the unit runs slightly warm to the touch - never hot.) If necessary, schedule compressor for adjustment of the clearance between the rotor and cones.

3. Listen to both check valves at the pump to determine if air leakage exists. Check condition of motor bearings.

4. Check line connections for air leaks.

5. Remove float valve assembly from separator; inspect and clean it. Reinstall and check water level on gauge glass; it should be 1/3 of the glass with the compressor operating.

Page 1 of 2

**Figure 4.44** PM Guide Number 111—Air compressor (centrifugal), low pressure. (*From Georgetown University.*)

**GEORGETOWN UNIVERSITY PREVENTIVE MAINTENANCE(GUIDE NUMBER)** 111
-continued

6. Clean the strainers in seal water line of each pump.

7. Check packing gland to ensure that leak off is only a slight drip. If necessary, replace packing seven rings of graphite impregnated packing, 5/16". Ensure that leak-off drain line is unobstructed.

**TOOLS AND MATERIALS:** Revision Date: 11/3/88

**Figure 4.44** *(Continued)*

**GEORGETOWN UNIVERSITY PREVENTIVE MAINTENANCE(GUIDE NUMBER)** 112

**EQUIPMENT:** Vacuum-Heating / Condensate Pump (St. Mary's)

**MANUFACTURER:**

**CATEGORY:** **FREQUENCY:**

**PREVENTIVE MAINTENANCE DESCRIPTION:**

1. Conduct annual maintenance of Vacuum Heating Pump.

**SPECIAL INSTRUCTIONS:**

**PROCEDURE:**

1. Check contacts of float switches and ensure that switches operate condensate pumps. Check "lead" and "lag" operation of switches. Check motor bearings.

2. Check to see if vacuum pump is handling air properly: hold the vacuum relief valve in by hand, or remove the pipe plug in the vacuum control line; air should be sucked in by the pump. Ensure vacuum relief valve seats properly.

3. Check separator float valve and vacuum pump seal water orifice for condition by observing drain line opening. When pump is not running, nothing should come out. When pump is running, only air, with no water, should be coming out.

4. Remove strainer flange and remove strainer for inspection; clean, if necessary. If stainer contains sludge that cannot be removed otherwise, use solvent for removal; do not damage strainer by hammering in order to clean it.

5. See that the air discharge is open to the atmosphere and that the check valve in the suction line to each vacuum pump is working properly. Access to the check valve is obtained by removing the check valve cover on the side of the tank adjacent to vacuum pump.

6. Lube motors.

**TOOLS AND MATERIALS:** **Revision Date:** 11/3/88

**Figure 4.45** PM Guide Number 112—Vacuum heating (condensate pump). (*From Georgetown University.*)

GEORGETOWN UNIVERSITY PREVENTIVE MAINTENANCE(GUIDE NUMBER) 113

EQUIPMENT: Air Compressor (Centrifugal) High Pressure

MANUFACTURER:

CATEGORY: FREQUENCY:

PREVENTIVE MAINTENANCE DESCRIPTION:

1. Conduct annual maintenance check on HP Nash air compressor.

SPECIAL INSTRUCTIONS:

PROCEDURE:

1. Lift pressure relief valve of air receiver tank and hold open. Observe pressure gauge to see of pump cuts in at the prescribed pressure and then pumps up tank under usage of air. Check cut-out pressure; release valve and check for proper seating. Check relief valve of separator by hand for proper operation. Prescribed cut-in / cut-out pressures:
Reiss Science: Lead cut-in at 65 psi; Lead cut-out 80 psi
Lag cut-in psi; Lag cut-out psi
When pump alternating capability exists, ensure monthly lead-lag change.
2. If pump capacity is observed to drop off during the above pump-up test, check for proper amount of water seal and that the compressor suction is unobstructed. (Seal water should be supplied in sufficient quantity that the unit runs slightly warm to the touch - never hot.) If necessary, schedule compressor for adjustment of the clearance between the rotor and cones.
3. Listen to both check valves at the pump to determine if air leakage exists. Check condition of pump and motor bearings.
4. Check line connections for air leaks.
5. Remove float valve assembly from separator; inspect and clean it. Reinstall and check water level on gauge glass; it should be 1/3 of the glass with the compressor operating.
6. Clean the strainers in seal water line of each pump and the recirculating line to the separator.
7. Check mechanical seal for excessive leak-off.
8. Replenish oil supply of the drive-end bearing as needed with oil.

TOOLS AND MATERIALS: Revision Date: 11/03/88

**Figure 4.46** PM Guide Number 113—Air compressor (centrifugal), high pressure. (*From Georgetown University.*)

GEORGETOWN UNIVERSITY PREVENTIVE MAINTENANCE(GUIDE NUMBER) 116

EQUIPMENT: Fire System Control Valves

MANUFACTURER: All

CATEGORY: Safety FREQUENCY: Yearly

PREVENTIVE MAINTENANCE DESCRIPTION:

Lubricate valve stems.(Fire Suppression Systems)

SPECIAL INSTRUCTIONS:
Always lock the valve in the open position before leaving the work site. (For any length of time!)

PROCEDURE:

A. Oil or grease valve stem completely. Close and reopen the valve to distribute the lubricant. Lock or seal the valve in the open position.

B. Check and if necessary adjust valve packing gland to stop any leakage. Report any leaks that cannot be repaired during this PM.

TOOLS AND MATERIALS:
Lubricant for valve stem. Hand tools to adjust packing gland.

**Figure 4.47** PM Guide Number 116—Fire system control valves. (*From Georgetown University.*)

GEORGETOWN UNIVERSITY PREVENTIVE MAINTENANCE(GUIDE NUMBER) 119

EQUIPMENT: Life Line T AC Motors 143T-449T

MANUFACTURER: Westinghouse

CATEGORY: Electric Motors FREQUENCY: Yearly

PREVENTIVE MAINTENANCE DESCRIPTION:

Westinghouse AC Motor Maintenance

SPECIAL INSTRUCTIONS:

OSHA Standard 1910.213 -Open lock and tag motor disconnect switch before starting work.

PROCEDURE:

**Motor Location**

Open Drip Proof motors are intended for relatively clean dry atmospheres of 104F ambient temperature or less. There are several options offered by Westinghouse that allow use of the motor at higher temperatures. See the modification Data sheets in the back of the PM Guide Book.

**Electrical**

Check insulation resistance periodically. Any approved method of measuring insulation resistance may be used provided the voltage across the insulation is at a safe value for the type and condition of the insulation. A hand cranked megger of not over 500 volts is the most convenient and safest method.

The recommended insulation resistance of stator windings tested at operating temperature should not be less than:

$$\frac{\text{rated voltage of machine} + 1000}{1000} = \text{Insulation resistance in megohms}$$

If the insulation fails to meet the calculated value then it will have to be dried out.

**Motor Lubrication**

Lubricating grease should be suitable for an operating ambient temperature of -15F to 130F. All Westinghouse motors are designed to be greased with the relief plug removed. Run the motor for 10 minutes before replacing plug.

Pg 1 of 2

**Figure 4.48** PM Guide Number 119—Life line T AC motors, 143T–449T. (*From Georgetown University.*)

GEORGETOWN UNIVERSITY PREVENTIVE MAINTENANCE(GUIDE NUMBER) 119

| Type of Enclosure | Insulation | Frame Size 143 to 215T | 254 to 326 | 364 to 449T |
|---|---|---|---|---|
| open-DP | B | 2 Yrs | 18 Mo. | 1 Yr |
| enclosed-FC<br>open-DP | B<br>F | 18 Mo. | 1 Yr | 9 Mo. |
| enclosed-NV<br>enclosed-FC<br>open-DP<br>enclosed-Lint free-FC | B<br>F<br>H<br>B | 1 Yr. | 9 Mo. | 6 Mo. |
| enclosed-NV<br>enclosed-FC<br>enclosed-Lint free-FC | F<br>H<br>F | 9 Mo. | 6 Mo. | 3 Mo. |

NOTES* -For motors over 1800 rpm use 1/2 of tabled period
-For heavy duty-dusty locations use 1/2 of tabled period
-For severe duty-high vibration, shock use 1/3 of table.

Volume-Reference Table

| shaft diameter (at face of bracket) | amount of grease to add |
|---|---|
| 3/4"to 1.25" | .1 oz |
| 1.25" to 1-7/8" | .2 oz |
| 1-7/8" to 2-3/8" | .6 oz |
| 2-3/8" to 3-3/8" | 1.6 oz |

Oil Lubricated Sleeve Bearings

Before starting the motor, fill both reservoirs through the filler plug with best quality, clean motor oil. The oil should have a viscosity of from 180 to 200SSU (equivalent to SAE #10). During operation, no oil should be added until it drops below the full level. Do not flood the bearing. At about 2 year intervals, dismantle and thoroughly wash out the bearing housing, using hot kerosene oil.

**TOOLS AND MATERIALS:** **Revision Date:** 8/4/88
#2 grease with an operating ambient temperature of -15F to 130F, grease gun, hand tools, rags for clean up.

pg 2 of 2

**Figure 4.48** (*Continued*)

GEORGETOWN UNIVERSITY PREVENTIVE MAINTENANCE(GUIDE NUMBER) 121

EQUIPMENT: Tri Clad 55 143 to 215 143T to 215T

MANUFACTURER: General Electric (G.E.)

CATEGORY: Electric Motors FREQUENCY: Yearly

PREVENTIVE MAINTENANCE DESCRIPTION:
General Electric AC motor maintenance

SPECIAL INSTRUCTIONS:
OSHA Standard 1910.213 -Open lock and tag motor disconnect switch and starter before starting work. Check the voltage of all incoming line terminals. Positively ascertain that the equipment is deenergized.

PROCEDURE: INSTALLATION

1. Location.
   a. **Dripproof Motors** are designed for installation in a well ventilated place where the atmosphere is reasonably free of dirt and moisture.
   b. **Standard Enclosed Motors** are designed for installation where motor may be exposed to dirt, moisture and most outdoor conditions.
2. Belts.
   a. Sheave ratios greater than 5:1 and center-to-center distances less than the diameter of the large sheave should be referred to the Company.
   b. **Tighten belts** only enough to prevent slippage. Belt speed should not exceed 5000 ft. per min.
   c. **V-Belt Sheave Pitch Diameters** should not be less than the following values:

| Horsepower | | | | V-Belt Sheave, Min. Dia. | |
|---|---|---|---|---|---|
| Synchronous Speed, Rpm | | | | Conventional* A and B Pitch Dia. | Super♠ 3V Outside Dia. |
| 3600 | 1800 | 1200 | 900 | | |
| 1½ | 1 | 3/4 | ½ | 2.2 | 2.2 |
| 2-3 | 1½-2 | 1 | 3/4 | 2.4 | 2.4 |
| --- | 3 | 1½ | 1 | 2.4 | 2.4 |
| --- | --- | 2 | 1½ | 2.4 | 2.4 |
| 5 | --- | --- | --- | 2.6 | 2.4 |
| 7½ | 5 | --- | --- | 3.0 | 3.0 |
| 10 | 7½ | 3 | 2 | 3.0 | 3.0 |
| --- | --- | 5 | 3 | 3.0 | 3.0 |
| 15 | 10 | --- | --- | 3.8 | 3.8 |

*Max sheave width=2 (N-W)-¼"
♠Max sheave width=N.W.

**Figure 4.49** PM Guide Number 121—Tri clad 55, 143T–215T (electric motors). (*From Georgetown University.*)

**GEORGETOWN UNIVERSITY PREVENTIVE MAINTENANCE(GUIDE NUMBER)** 121
(continued)

| Frame Size | | Min |
|---|---|---|
| 1 Phase | 3 Phase | Sheave Diameter |
| 182<br>184,213,215 | 182<br>184,213<br>215 | 2¼"<br>2½"<br>3" |

3. Motor Windings

a. To clean, use a soft brush and, if necessary, a slow acting solvent in a well ventilated room. In drying, do not exceed 85°C(185°F).

4. Inspection

a. **Inspect** motor at regular intervals. Keep motor clean and ventilating openings clear.

5. Lubrication

a. **Ball-Bearing Motors** are adequately lubricated at the factory. Relubrication at intervals consistent with the type of service (see table) will provide maximum bearing life. Excessive or too frequent lubrication may damage the motor.

b. **Motors having pipe plugs or grease fittings in bearing housings** should be relubricated while warm and at stand-still. Replace one pipe plug on each end shield with 1/8" inch pipe thread lubrication fitting. Remove the other plug for grease relief. Be sure fittings are clean and free from dirt. Using a low-pressure grease gun, pump in the recommended grease until new grease appears at grease-relief hole. After relubricating, allow motor to run for 10 minutes before replacing relief plugs.

c. **Motors not having pipe plugs or grease fittings in bearing housings** can be relubricated by removing end shields from motor, cleaning grease cavity and refilling 3/4 of circumference of cavity with recommended grease.

**Caution:** Bearings and grease must be kept free of dirt.

Pg 2 of 3

**Figure 4.49** *(Continued)*

**GEORGETOWN UNIVERSITY PREVENTIVE MAINTENANCE(GUIDE NUMBER)** 121
(cont)

| Type of Service | Typical Examples | Hp Range | Relubrication Interval |
|---|---|---|---|
| Easy | Valves; Door Openers; portable floor sanders, motors operating infrequently | ½-7½<br>10-40<br>50-150<br>200-250 | 10 Years<br>7 Years<br>4 Years<br>3 Years |
| Standard | Machine tools; air-conditioning apparatus; conveyors; garage compressors; refrigeration + laundry machines; water pumps,woodworking | 1½-7½<br>10-40<br>50-150<br>200-250 | 7 Years<br>4 Years<br>1½ Years<br>1 Year |
| Severe | Motors for fans; M-G sets; (running 24 hours/day, 365 days/year; motors subject to severe vibration,etc. | 1½-7½<br>10-40<br>50-150<br>200-250 | 4 Years<br>1½ Years<br>9 Months<br>6 Months |
| Very Severe | Dirty, vibrating applications; where end of shaft is hot (pumps and fans); high ambient | 1½-7½<br>10-150<br>200-250 | 9 Months<br>4 Months<br>3 Months |

**TOOLS AND MATERIALS:** **Revision Date:**8/16/88
#2 grease with an operating ambient temperature of -15F to 130F, grease gun, hand tools, rags for clean up.

Pg 3 of 3

**Figure 4.49** *(Continued)*

GEORGETOWN UNIVERSITY PREVENTIVE MAINTENANCE(GUIDE NUMBER) 125

EQUIPMENT: Rotary Vacuum Pump

MANUFACTURER: Leiman Bros. / ITT

CATEGORY: Lubrication FREQUENCY: Monthly

PREVENTIVE MAINTENANCE DESCRIPTION:

Change Oil on Rotary Vacuum Pump (Leiman/ITT)

SPECIAL INSTRUCTIONS:

OSHA Standard 1910.213 -Open lock and tag motor disconnect switch before starting work.

PROCEDURE:

A) Drain old oil and flush tank with new oil.

B) Refill the automatic oil feed tank to the top of the oil fill elbow with SAE 30 ND oil.

C) With pump running at required vacuum adjust oil valve to 2 to 3 drops per minute on all pumps except 100 and 107 which should be 6 drops per minute.

D) Make sure that gasket to oil tank is sealed properly and there are no air leaks.

TOOLS AND MATERIALS: Revision Date: 8/23/88

SAE 30 non detergent oil, container to discard old oil, hand tools to remove covers.

**Figure 4.50** PM Guide Number 125—Rotary vacuum pump. (*From Georgetown University.*)

## Energy Management Control System (EMCS)

The EMCS is used to control mechanical and electrical systems that provide required environmental interior temperatures in buildings. The EMCS is a computer-based system featuring a microprocessor that starts, stops, and monitors mechanical and electrical systems, and their individual components, throughout the building on a continuous or scheduled basis.

### Operation

The EMCS should be operated from the Work Control Center. This operation should include surveillance of the building rooms, areas, and mechanical systems for adherence to prescribed environmental temperatures. The EMCS operator should take corrective actions to maintain the necessary environmental temperatures within buildings by making any needed EMCS adjustments on a daily basis.

#### Emergency or unscheduled EMCS shutdown

The FD should plan to have on call additional temporary personnel, or contractor support personnel, necessary to maintain full and acceptable performance from all building systems during any EMCS failure. This plan should become effective immediately and at such time as building environmental temperatures cannot be maintained using manual adjustments to the building systems. This plan should remain in effect until such time as the EMCS becomes fully functional as originally designed.

## Equipment and System Warranties

It is advisable to develop a computerized maintenance management system (CMMS) module program to track and manage all building equipment and system components that are under current valid warranties. This will assist in ensuring that required repairs or corrections, during the warranty periods, will be made by the installing contractors. Should there be a dispute as to the responsibility for the repair/adjustment, the facility manager will make a determination as to the responsible party and issue directions on how to proceed. The FD should provide operational service call response (and corrective action), and preventive maintenance functions to the deficient item or system even though the item is under a manufacturer's warranty and may be awaiting repairs or adjustments by others.

## PREDICTIVE MAINTENANCE PROCEDURES

### Introduction

*Predictive maintenance* is a maintenance philosophy where equipment condition is monitored at appropriate intervals to enable accurate evaluation to use as input when determining whether maintenance action or "no action" is required without sacrificing equipment reliability.

Effective, reliable operation of a facility dictates that the facility's availability and intended functionality be optimized. One aspect of the optimization process is that of designing, implementing, and following through with an effective predictive maintenance program. Facilities maintenance has evolved from a "breakdown" (if it's not broken, don't fix it) philosophy in the 1950s and early 1960s to a *preventive maintenance* (time-based) philosophy in the 1970s and progressing to a *predictive* philosophy at present. This section is a summary of some of the most effective *predictive maintenance* technologies and processes which can help to optimize the availability, reliability, and profitability of both large and small facilities.

#### Benefits

Knowing the condition of your equipment at any point provides valuable information that can then be used as input to:

1. Anticipate the need for repairs by identifying faults in the early stages of development.

2. Order required parts only when needed to avoid stocking unnecessary inventory.
3. Schedule appropriate resources for repairs well in advance to minimize the impact on the process.
4. Eliminate unnecessary preventive maintenance tasks.
5. Avoid catastrophic failures and minimize additional damage which may occur.

With these benefits in mind, it is easy to understand how a formal, comprehensive, and effective predictive maintenance program will provide cost savings that will improve the profitability of any facility. The actual savings realized by a particular facility will vary based on the size of the facility, the quantity and type of equipment involved, and the extent and effectiveness of the program.

### Technologies

There are many technologies or "tools" of predictive maintenance available that aid in predicting the need for maintenance action or no action. Perhaps the simplest tool is that of the operating and maintenance personnel. They know the equipment and are in close proximity to the equipment on a daily basis. Therefore, their eyes and ears are capable of detecting very subtle changes in the operating condition of the equipment. As such, they should be directed and encouraged to be alert for these changes and, when changes are detected, act to have the equipment's condition analyzed further for abnormalities. When looking past the personnel for predictive maintenance tools, four specific technologies stand out as the most effective for application in most facilities. These are:

- Increased frequency of monitoring
- Additional testing (temperature, oil, etc.)
- Scheduled inspection or repairs
- Immediate shutdown to avoid catastrophic failure

The overall levels of vibration which are considered "acceptable" can be obtained from a number of sources including:

- Equipment manufacturer's recommendations
- Industry guidelines
- Historical levels of like or similar machines

Figure 4.51 is one of many charts published to provide a guide to operating and maintenance personnel to help determine the overall condition of their machinery. However, it is just a snapshot look at vibration severity and must be used only as a guide.

Once it is determined that the vibration on a machine exceeds the acceptable limits, additional analysis must be performed to determine the source. This is accomplished by identifying the frequencies of the vibration generated by the machine. The specific frequencies can be related to specific faults and components of the machine. For instance, for a machine that operates at 3600 rpm (60 Hz), any unbalance of the machine's rotor will be represented by the existence of vibration at a frequency of 60 Hz. The "unbalance" may be the result of wear, loss of material, or material buildup. Once identified as unbalance, steps can be taken to reduce the unacceptable levels of vibration by balancing the rotor, repairing worn parts, or cleaning the buildup from the rotor. Similarly, a looseness in the same machine or its mounting may exhibit a frequency of 120 Hz, or two times its rotating speed, as a result of the rocking motion resulting from looseness. Figures 4.52 and 4.53 show typical plots of vibration amplitude versus frequency, one of the more effective displays of a machine's characteristics used in the diagnostic process.

Also, many machine faults can be identified by identifying the frequency, direction, phase relationship, and other characteristics of the vibration exhibited by the machine. Table 4.1 shows the relationships of various characteristics of vibration to specific machinery faults. It should be noted that the relationships are based on probabilities and are not intended to be hard and fast rules. Machinery analysts are continually finding machines which exhibit unusual vibration characteristics which challenge their analytical skills when they fail to fit the generally accepted rules of diagnostics.

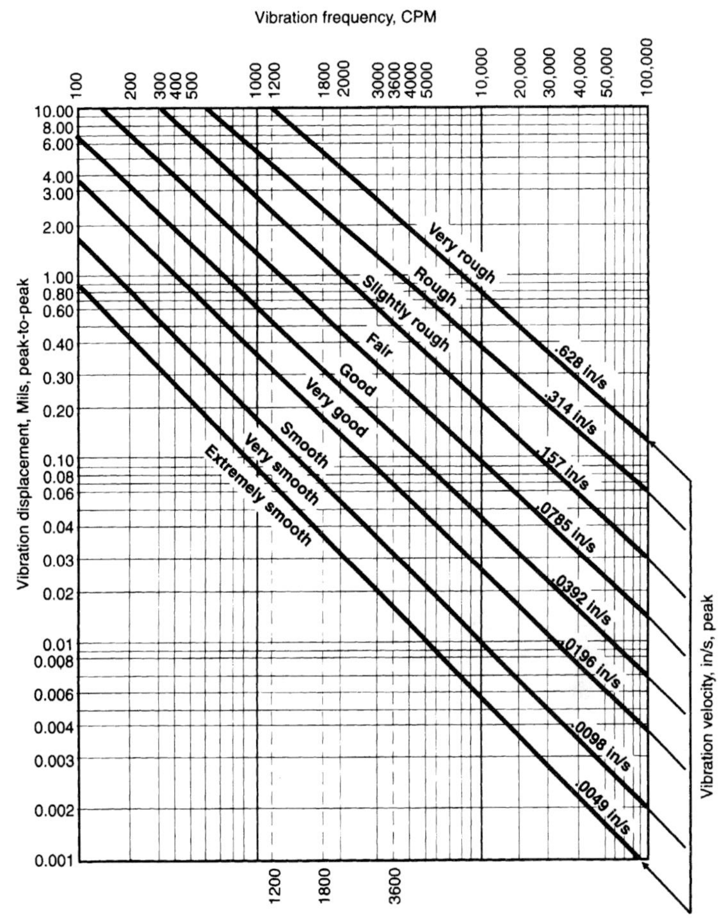


**Figure 4.51** Vibration severity chart. Readings shown are for filtered readings taken on the machine structure or bearing cap. (*From IRD Training Manual, General Machinery Vibration Severity Chart, Form #3050, IRD Mechanalysis, Columbus, Ohio.*)

In order to have an effective program, the vibration exhibited by a machine must be trended over time. To accomplish this there is a considerable range of vibration data loggers on the market. The basic instrumentation will display, in analog form, overall vibration levels which can then be recorded and trended manually. The latest microprocessor based data collectors are very powerful and function as fully functional

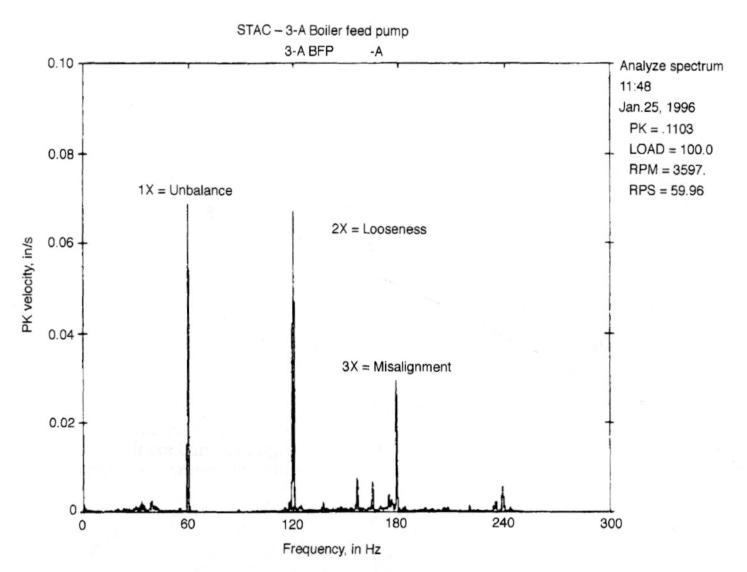


**Figure 4.52** Vibration amplitude versus frequency.

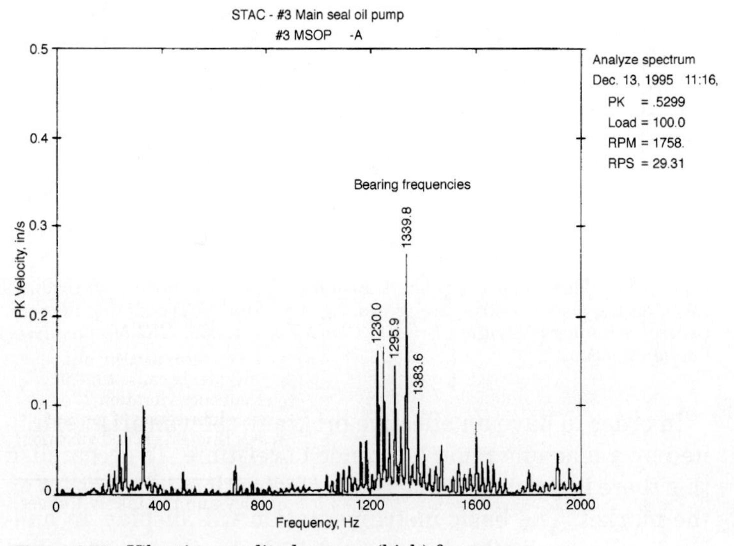


**Figure 4.53** Vibration amplitude versus (high) frequency.

**TABLE 4.1 Vibration Frequency and the Likely Causes**

| Frequency in terms of rpm | Most likely causes | Other possible causes and remarks |
|---|---|---|
| 1 × rpm | Unbalance | 1. Eccentric journals, gears, or pulleys<br>2. Misalignment or bent shaft—if high axial vibration<br>3. Bad belts if rpm of belt<br>4. Resonance<br>5. Reciprocating forces<br>6. Electrical problems |
| 2 × rpm | Mechanical looseness | 1. Misalignment if high axial vibration<br>2. Reciprocating forces<br>3. Resonance<br>4. Bad belts if 2 × rpm of belt |
| 3 × rpm | Misalignment | Usually a combination of misalignment and axial excessive clearances (looseness) |
| Less than 1 × rpm | Oil whirl (less than $^1/_2$ rpm) | 1. Bad drive belts<br>2. Background vibration<br>3. Subharmonic resonance<br>4. "Beat" vibration |
| Synchronous (ac line frequency) | Electrical problems | Common electrical problems include broken rotor bars, eccentric rotor, unbalanced phases in polyphase systems, unequal air gap |
| 2 × synchronous frequency | Torque pulses | Rare as a problem unless resonance is excited |
| Many times rpm (harmonically related frequency) | Bad gears<br>Aerodynamic forces<br>Hydraulic forces<br>Mechanical looseness<br>Reciprocating forces | Gear teeth times rpm of bad gear<br>Number of fan blades times rpm<br>Number of impeller vanes times rpm<br>May occur at 2, 3, 4, and sometimeshigher harmonics if severe looseness |
| High frequency (not harmonically related) | Bad antifriction bearings | 1. Bearing vibration may be unsteady—amplitude and frequency<br>2. Cavitation, recirculation, and flow turbulence cause random high-frequency vibration<br>3. Improper lubrication of journal bearings (friction-excited vibration)<br>4. Rubbing |

SOURCE: *IRD Training Manual,* Vibration Frequency and the Likely Causes, IRD Mechanalysis, Columbus, Ohio.

Fast Fourier Transform (FFT) analyzers capable of recording frequency spectra, time waveform, and other relevant vibration data in various forms. This data is then stored, trended, viewed, and manipulated through standalone or networked PCs. The current software packages available for this purpose can be very comprehensive with the more advanced versions offering Expert Systems to aid in the diagnosis of the vibration data.

Regardless of the particular hardware and software package chosen, vibration monitoring is an imperative predictive tool for any facility which relies on rotating machinery. Many vibration monitoring programs in a wide range of industries have recovered their costs in less than one year.

## Infrared (IR) Thermography

In the last several years monitoring equipment temperatures using infrared measurement technology has become a very effective predictive maintenance tool. One reason for this is that the cost of the measurement instrumentation has in some cases become more reasonable while the performance has improved considerably. Also, we have learned that most equipment experiences an increase in temperature as faults develop. Identifying those faults early in their inception provides valuable time for planning repairs, procurement of necessary parts, and allocating resources with minimum disruption of the facility's process.

The measurement instrumentation may be a simple infrared measurement device that can be interfaced with the same data logged used for vibration measurement. Measurements such as machine bearing temperature, cooler inlet and outlet temperatures, motor air inlet and outlet temperatures, etc., can be stored, trended, and automatically alarmed using the same computer hardware and software used with the vibration monitoring portion of the predictive program. Figure 4.54 is an example of a trend of temperature using this method.

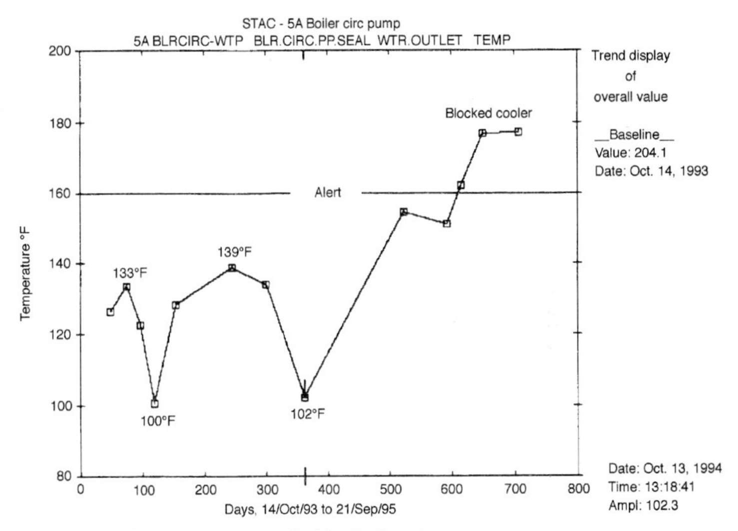


**Figure 4.54** Temperature trend—blocked cooler.

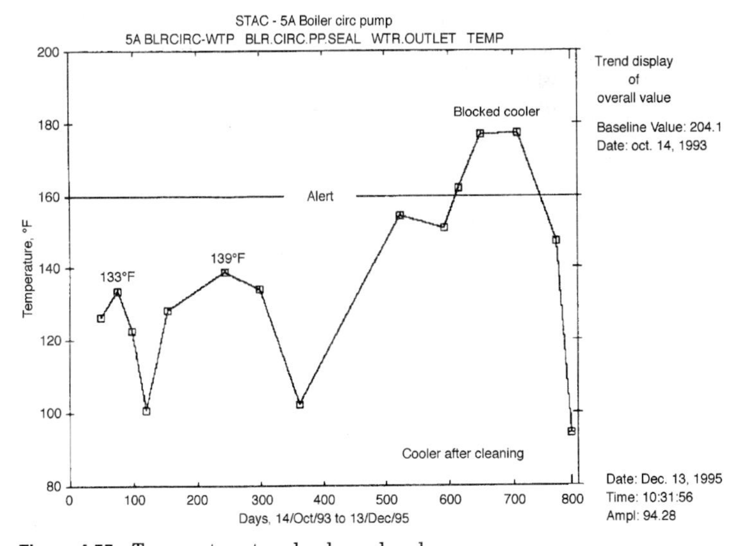


**Figure 4.55** Temperature trend—cleaned cooler.

Figure 4.55 indicates the resulting reduction in temperature after the cooler was found to be partially plugged and subsequently cleaned. This action prevented serious damage to the machine.

More sophisticated thermal imaging cameras are used as comprehensive analysis tools and are most effective for quick scanning of a large number of components for fault identification, as well as for applications where heating or cooling is a process variable.

Infrared imaging has proven particularly effective on:

- Motors
- Motor control centers
- Load centers
- Breakers
- Circuit board components
- Transformers
- High voltage disconnects
- Transmission and distribution systems

Figure 4.56 is an example of the type of electrical fault which is easily identifiable with a quick scan on an energized circuit. It is important to realize not only are the faults identified but during a typical thermal survey better than 90 percent of the equipment can have routine, unnecessary preventive maintenance deferred. No longer will the periodic, time-consuming task of checking each connection for corrosion and looseness be necessary. As such, only the maintenance that is required will be identified and accomplished making effective use of resources.

In order to classify the severity of electrical heating anomalies, guidelines have been established by various institutions such as the Electric Power Research Institute (EPRI) to aid in the planning and scheduling of repairs. Table 4.2 is one example of these guidelines.

In addition to the electrical applications already mentioned, IR thermography can be effectively applied to the following:

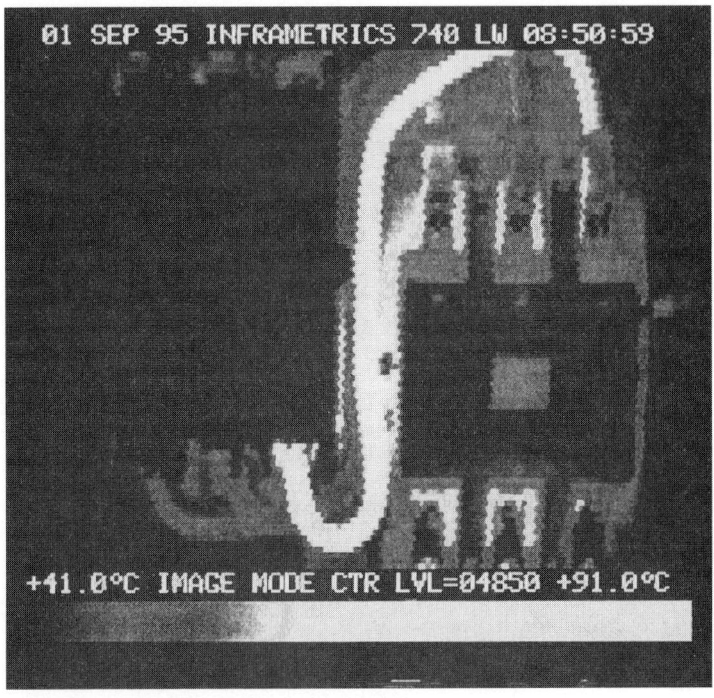


**Figure 4.56** Thermal image of electrical fault.

**TABLE 4.2 Temperature Rise Classification Guide**

| | | |
|---|---|---|
| Minor problem | 1–10°C rise | Increase monitoring frequency |
| Intermediate problem | 10–20°C rise | Repair within one year |
| Serious problem | 20–30°C rise | Repair in near future |
| Critical problem | 30°C or greater rise | Repair immediately |

1. Insulation degradation
   - Building siding and roofing
   - Steam piping
   - Heated and refrigerated spaces, tanks, etc.
   - Boiler and furnace walls
   - Electric motor hot spots

2. Steam traps
3. Leaking valves
4. Oil, water, and other liquid tank levels
5. Below grade piping leaks and ruptures
6. Rotating machinery faults
   - Seal rubs
   - Faulty bearings
   - Coupling misalignment
   - Misaligned belts
   - Uneven heating/cooling due to blockages
7. Restricted flow due to blockages
   - Water or fluid lines
   - Coal supply lines
   - Boiler tubes
8. Process applications
   - Mass production utilizing a heating/cooling process
   - Bonding/lamination processes
9. HVAC
   - Leaks, blockages, restrictions
   - Proper heating and cooling distribution
   - Compressors, fans, and motors

Routine temperature monitoring at regular intervals is a necessary aspect any facility's predictive maintenance program. However, the extent and frequency of monitoring will vary with the equipment monitored and its importance to the process. For instance, monitoring and trending the temperature of the bearings of the critical rotating machines weekly or monthly is reasonable. Using a thermal imaging camera to scan all motor control centers and breakers may only be required on an annual basis. Again, these intervals will vary and should be defined during the initial setup of the predictive program and reviewed and revised as needed.

## Oil analysis

Effective monitoring of equipment condition includes development and implementation of a formal lubricant analysis program. The program serves two major functions:

- Identifying lubricant condition
- Identifying equipment wear

Knowing the condition of the machinery lubricant provides valuable knowledge which can be used as an integral part of a facility's overall maintenance program. Many facilities have machinery which require large quantities of lubricants. These lubricants are periodically replaced even if, based on their condition, they do not require replacement. When the oil is removed and discarded, valuable information about the wear of the components may be discarded with the oil. Each time the lubricant is replaced, environmental concerns must also be considered.

## Lubricant condition

Monitoring oil to determine its condition includes a look at contamination from water, fuel, dirt, and other substances. Knowing the type and quantity of contaminants will often lead to identification of other machinery problems which can then be addressed. The lubricants have been chosen by the equipment manufacturer for their particular properties for each application. Some of these properties are achieved by additives. If a facility is interested in deferring unnecessary oil changes, knowing the oil has the desired properties is imperative. Therefore, appropriate tests are performed to monitor these properties.

Many oil-monitoring programs, particularly during the initial sampling, have identified improper oil grade or type for a particular machine. This condition could explain premature failures and the need for unnecessary repairs. This discovery alone may be justification for initiating oil monitoring.

### Equipment wear

The predictive maintenance process is aimed at predicting when maintenance is needed and when it is not. While vibration monitoring is certainly the most widely used tool for determining rotating machinery condition, oil analysis will, in many situations, provide an earlier indication that abnormal or premature wear is in progress. Oil monitoring and analysis is especially appropriate for slow speed machines, reciprocating machines, and gear boxes, as they usually show developing faults earlier using oil versus vibration analysis. As internal machine components wear, they leave the wear particles in the lubricating oil. Identifying the existence, size, shape, and elements of the wear particles leads to identifying the particular component experiencing the wear. This valuable information can then be used to aid in determining the ability of the machine to continue operating, plan for repairs, order necessary parts, and prevent unnecessary, unplanned downtime. The wear may be the result of many mechanisms including improper maintenance practices, misalignment, overheating, improper lubrication, improper operation, or poor design application. By monitoring and trending the wear particles, initial levels can be established and changes can be recognized. Figure 4.57 illustrates the typical wear pattern of machine components over time.

The initial break-in period will indicate an increase in wear particles as the parts wear in and the particle concentration stabilizes. With continued monitoring at appropriate intervals, excessive wear will become evident by an increase in the wear particle concentration. At that time a comprehensive analysis may be initiated to determine the particle type, wear mechanism, and the source.

### Oil testing

Numerous tests and instruments have been developed to provide accurate information about the condition of machinery lubricants. Initial testing should be a screening process to identify those cases which require additional assessment

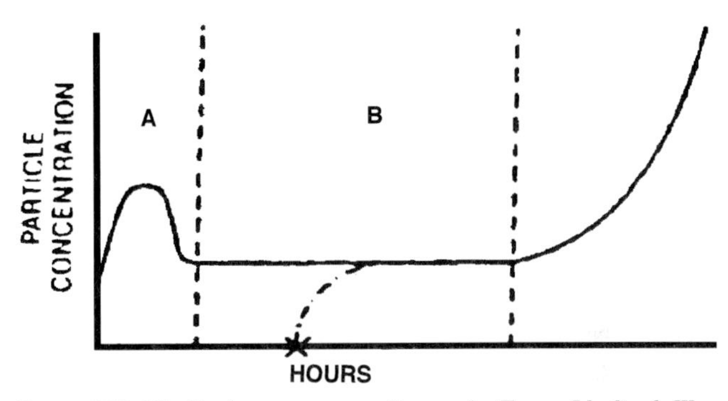


**Figure 4.57** Idealized wear curve. (*Larry A. Toms, Idealized Wear Curve, Machinery Oil Analysis, Methods, Automation and Benefits, 1995, p. 34, Fig. 3.4.*)

and to eliminate the need to further evaluate the vast majority of machines which are operating properly. As with any predictive technique, each machine should be evaluated to determine which tests are appropriate and to assign the appropriate monitoring frequency.

Sampling methods have a great impact on the results of the oil analysis program. The sampling is where the analysis process begins and if not done properly will affect the results including the possibility of performing unnecessary maintenance. Formal sampling guidelines should be established and adhered to. Many of the oil analysis labs will provide sampling guidelines for the program.

The specific tests which are determined to be appropriate and cost effective can vary considerably depending on the particular machine, its service and environment, the type of lubrication used, and the laboratory performing the testing. Some large facilities may have the resources to perform some or all of the tests in-house. However, most facilities find that contracting the testing from a credible laboratory is effective, accurate, and reasonably priced. Many of the modern labs have mechanized testing devices which allow for large quantities of samples to be processed each day while maintaining very competitive costs. Also, some labs provide the sample results in electronic form which can be downloaded to a facilities predictive maintenance database for trending, analysis, correlation with other predictive

**TABLE 4.3 Lubricant Condition Tests**

| Test parameters | Test options | Comments |
|---|---|---|
| Viscosity | ASTM D445 | Measures oil's resistance to flow |
| Oxidation | Total acid number (TAN) ASTM D974/D664 | Determines acidity level |
| | Total base number (ASTM D2896) | Determines alkalinity level. Used to detect fuel or coolant contamination |
| | Fourier transform infrared analysis (FTIR) | Screening test only. Exceptions require TAN test |
| Water | Appearance | Visual check |
| | Karl Fischer reagent (ASTM D1744) | Gives water content in ppm. Accurate to 10% |
| | Fourier transform infrared analysis | Exceptions require Karl Fischer reagent test |
| | Crackle test | Oil on a hot plate. Subjective test |
| Solids | Light extinction | Oil passed through photodetectors. Particles are counted and classified by size |
| | Mesh obscuration | Oil forced through different-sized meshes and counted under microscope |

technologies, and reporting. Table 4.3 shows some of the tests that may be used to determine the condition of the lubricant. Table 4.4 indicates some of the more common tests performed to identify wear particles used for determining machine condition.

These are only some of the many tests available for analyzing oil condition. There are also a variety of opinions as to which tests are the best. Discussions with suppliers of the services and instrumentation will provide insight into which combination of tests are appropriate for a particular facility.

**TABLE 4.4 Machine Wear Tests**

| Wear particle test | Test description |
|---|---|
| Spectroscopic analysis | Determines concentration (ppm) levels of key elements in oil |
| Direct ferrography | Provides large and small ferrous particle count |
| Analytical ferrography | Quantifies extent, type, and distinguishing features of particles |

Although most suppliers of lubricating oils are reputable, a facility's oil-monitoring program can also be used to provide a means of acceptance testing on deliveries of new lubricating oils. This also provides a baseline of values for comparing future samples drawn from the machinery.

Knowing the condition of the lubricating oils provides valuable information which can be used as one of the many facets of a facility's predictive maintenance program. When properly developed, implemented, and periodically updated, it will be an effective tool for controlling maintenance costs.

### Ultrasound

High-frequency sound is generated by many mechanical and electrical systems. This sound is not audible to the human ear, which hears in a frequency range up to 20,000 Hz. Instead, the high-frequency, low-level sound generated by the early stages of improperly lubricated bearings, leaking valves, and other faults appears in the frequency range above 20,000 Hz. To detect these faults early in their development, monitoring instrumentation has been developed which detects and converts this high-frequency sound into the audible range for detection by the analyst with headphones and for recording on data loggers and printers if desired.

The ultrasound frequency range used for fault detection is typically in the range of 20,000 to 100,000 Hz. This high frequency has a relatively short wavelength and, therefore, tends to travel in straight lines and over short distances. This characteristic provides an opportunity to use ultrasonic detection to zero in on particular faults, such as an improperly lubricated bearing, a leaking valve, or leaks in pressurized systems. Identifying the exact component early in its deterioration will provide an opportunity to plan for repairs, order necessary parts, and schedule resources with minimum impact on a facility's process.

Ultrasonic detection can be useful for airborne noise as well as contact measurements. In either application the filtering of sound below approximately 20 kHz effectively blocks out the typical machinery background noise allowing concentration on fault frequencies. Although different faults generate typical frequencies, the specific frequencies may vary with varying situations. Table 4.5 can be used as a guide to help identify some of the typical faults and the frequencies they generate.

### Ultrasound applications

As seen on the preceding chart, there are a number of applications for this technology. One very effective and widely used application is identification of leaks in pressurized systems. A system under pressure that develops a leak to atmosphere will generate high-frequency noise due to the expansion of the gas or liquid moving through an orifice from a high-pressure to a low-pressure environment. Similarly, vacuum leaks may also be detected although this is usually more challenging.

Detecting leaking valves is also an effective application of ultrasonic technology. If the system containing the valve is pressurized, a closed, leaking valve will also generate a high-frequency noise. With valve leak detection, using a contact probe is more appropriate than using airborne detection because it eliminates the background noise from the surronding area.

Antifriction (ball and roller) bearings in rotating machinery are also candidates for ultrasonic detection of developing faults. These bearings, even when new, generate high-frequency noise. However, when faults develop, the amplitude of the high frequency will increase. Based on

**TABLE 4.5 Ultrasound Frequency Chart**

| | Suggested module | | | | | | | | | | |
|---|---|---|---|---|---|---|---|---|---|---|---|
| | Fixed band | 20 kHz | 25 kHz | 28 kHz | 32 kHz | 40 kHz | 50 kHz | 60 kHz | 100 kHz | Meter mode | Selection |
| Steam traps | X | | X | | | X | | | | Log | Stethoscope |
| Valves | | X | | | | X | | | | Log | Stethoscope |
| Compressors (valves) | X | | X | | | | | X | | X | Log Stethoscope |
| Bearings | X | | | X | X | | | | | Lin | Stethoscope |
| Pressure and vacuum leaks | X | | | | | X | | | | Log | Scanner |
| Electrical arcs (and corona) | X | | | X | | | | | | Log | Scanner |
| Gears | | X | X | | | | | | | Log | Stethoscope |
| Pumps (cavitation) | X | X | X | | | | | | | Log | Stethoscope |
| Piping systems (underground) | X | X | | X | | | | | | Log | Stethoscope |
| Condenser tubes | X | | | | X | | | | | Log | Scanner |
| Heat exchangers (tone method) | X | | | | | | | | | Log | Scanner |

SOURCE: *Ultraprobe 2000 Instruction Manual,* Frequency Selection Chart, UE Systems, Elmsford, New York.

NASA research, the following gains in decibel level above baseline indicate various stages of bearing failure:

| | |
|---|---|
| 8 dB | Prefailure or lack of lubrication |
| 12 dB | Beginning of failure |
| 16 dB | Advanced failure |
| 35–50 dB | Catastrophic failure |

When electrical equipment begins to fail it may produce arching or corona. If heat is generated and the equipment is accessible, infrared thermography may be used to detect the problem. However, if there is no heat generated or the equipment is isolated, ultrasound can be an effective tool for fault identification. Both of these faults generate ionization and produce high-frequency airborne noise that is detectable with the ultrasound instrumentation.

Reciprocating compressors used for compressed air systems are often critical to the operation of a facility. Their unexpected failure is expensive in terms of both resources and lost production. Therefore, it is important to be able to identify potential problems with these machines. These machines have valves that must operate properly. It has been found that when these valves are leaking or sticking they exhibit a change in their sound pattern. This change can be detected using ultrasonic instrumentation. Figure 4.58 shows a comparison of a good valve and a failing valve using an ultrasound detector output providing input to a FFT analyzer.

Ultrasound is a relatively easy to use and inexpensive predictive tool that, for most facilities, will provide a very rapid return on investment.

### Integration

Although not actually a tool, integration of all the predictive maintenance technologies is a powerful diagnostic function. Also, integrating the predictive program with all the maintenance functions should be the goal of every facility.

Preventive maintenance (PM) has largely been a function where certain inspections and replacements have been performed at certain time intervals such as monthly, quarterly,

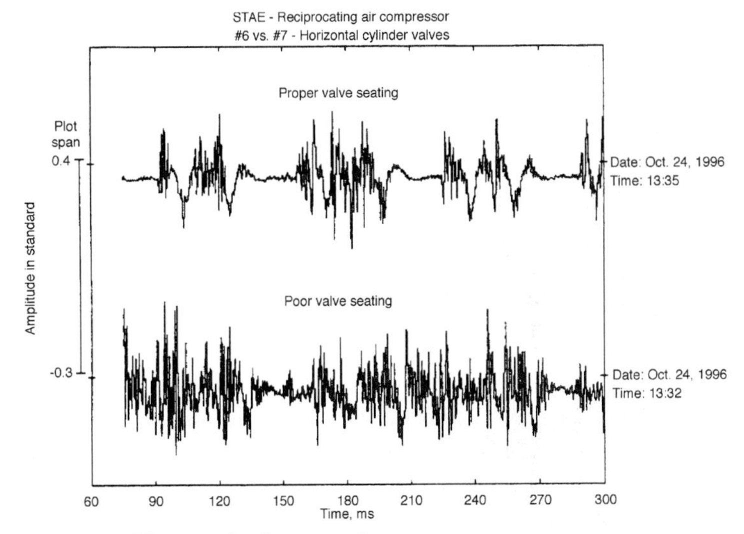


**Figure 4.58** Ultrasound valve comparison.

or yearly. Countless man-hours have been spent inspecting equipment that have no problems. Many machines have experienced failure following preventive maintenance action and may be referred to as "maintenance induced failure." If a facility's maintenance program were integrated, it would use the results of the predictive program to influence the need for preventive inspections and repairs. There are certainly justifiable reasons for performing time interval maintenance. However, integrating the condition of the machine, previous history, and manufacturers' recommendations would make the program more effective by eliminating unnecessary PM, reducing maintenance induced failures, and approaching "just in time" maintenance.

To illustrate the potential benefits of integrating a predictive maintenance program's technologies, consider a scenario where a large machine, critical to the process, is identified as experiencing an increase in the wear particle concentration in one of the bearing oil sumps. The recommendation from the lab is to increase the frequency of oil sampling. A subsequent oil sample shows another increase in wear particle count and, through analytical ferrography, identifies components of the bearing cage as the major contributor. A recommendation is made to shut down the machine for inspection of the bearing. With several days left in the production cycle, shutting down the machine now would result in a considerable loss of output. However, continuing to operate the machine in this condition may result in a catastrophic failure which would be even more costly in terms of repairs and down time. In this situation the use of vibration analysis and perhaps temperature monitoring may provide additional information which could be used to make a more informed decision regarding the course of action for this machine. For example, if the high-frequency vibration on this bearing showed a somewhat elevated level and, through a review of previous similar cases, it is known that, at those levels these bearings have operated for several weeks without failure, the course of action becomes more clear. At that time the frequency of monitoring vibration and temperature should be increased to several times a day until the production cycle is complete and repairs could be accomplished.

Knowing the condition of the facility's equipment by using and integrating all the predictive tools enables managers to make more informed maintenance decisions.

## Expert Systems

Expert Systems pertaining to predictive maintenance programs are currently under development with some fairly comprehensive packages available for vibration analysis. These systems still rely on interaction with a knowledgeable analyst; however, they do provide a means of speeding the process of reviewing large amounts of data. The rule-based Expert System is useful for providing suggestions for probable fault based on the information provided by the recorded data and by the machinery analyst. Vibration analysis is still considered to be an art as well as a science. This challenges the current Expert Systems and dictates the need for trained and experienced machinery analysts.

As the Expert Systems become more advanced and utilize information learned through the use of neural networks, they should prove to be a very effective tool for integrating and speeding the predictive maintenance process.

### Predictive process

Throughout the development of a predictive maintenance program it is important to remember the goal of the program. That goal is to provide the information needed to make decisions on the need for maintenance action or no action without sacrificing reliability. In order to achieve this goal, the program must be set up so that the proper parameters are monitored on the appropriate equipment at the appropriate frequency. There may be any number of approaches to the predictive process. What follows is one approach which will highlight some of the important considerations necessary in the development of the predictive maintenance program.

### Program startup

Once the decision is made to develop and implement a formal, integrated predictive maintenance program, the manager must ensure there is the needed support in the way of adequate resources to ensure the success of the program. Many predictive programs failed because they lacked management support. This is true for any program at any facility; without this support, it will fail. With that understood, the following steps need to be accomplished, though not necessarily in the same order. The extent to which each step is carried out will be somewhat different for each facility. It will depend on the size of the facility, the facility's process, and the level of commitment desired.

1. Select personnel
2. Identify equipment to be monitored
3. Select predictive technologies
4. Establish monitoring frequency
5. Select instrumentation (in-house applications)

6. Implement program
7. Establish baseline data
8. Set alert levels

### Select personnel

For the predictive maintenance program to succeed it must have dedicated, motivated personnel with the authority to perform the required functions. They must also be held accountable for the results of the program. These requirements are no different than those of any other program if the expectation is a success. Many predictive programs have failed because the personnel only perform the predictive maintenance function part-time when "things are slow" or when there is an emergency which requires immediate attention. Consider the manager of a facility who has other duties and only manages the facility part time! The results are in proportion to the effort. The same goes for the predictive maintenance personnel. This is not to say that the individual(s) cannot perform other functions. However, the program must be taken seriously in order to have a chance of success.

Another important factor in the success of the program is the training of the personnel. The entire field of predictive maintenance including its tools and techniques continues to evolve along with the development of other technological areas that depend on microprocessor based instrumentation and software. To keep informed and continually more effective, continued formal training is imperative. Fortunately many vendors are available to provide the required training. Many of the suppliers of the predictive instrumentation will provide, free of charge or for reasonable fees, valuable training specific to their tools.

### Identify equipment to be monitored

When selecting the equipment to be included in the monitoring program, a number of considerations need to be addressed. These include:

Criticality of the equipment

Cost to repair or replace

Cost of downtime

Past maintenance history

Manufacturers' recommendations

Program resources and limitations

The more critical the equipment to the continued operation of the process, the more seriously it should be considered for monitoring. If the failure of a piece of equipment reduces the facility's output, it is imperative to know the condition of the equipment and to know that it can be relied upon to provide continued service. Also, knowing the existence of a developing fault and the extent of the fault will provide information for scheduling repairs at the earliest opportunity without adversely affecting facility output—the goal of the predictive program.

The cost to repair or replace a piece of equipment should be one of the factors to consider when selecting candidates for the program. Equipment which has a long lead time for obtaining parts or replacement should be considered for the program. Generally, the greater the cost of the equipment, the greater the need to monitor. A comparison must be made between the cost to monitor and the cost of repair or replacement. For example, spending resources to monitor an inexpensive ventilation fan for bearing failure is not reasonable when the cost of the bearing repair or fan replacement is insignificant and it does not adversely affect the facility's process significantly. Let common sense prevail. Predictive maintenance is, after all, a common sense maintenance philosophy.

As mentioned above, if the equipment's failure causes down time for the process, the condition of the equipment should be known. This avoids surprises and reactionary maintenance which is almost always the most expensive.

A machine's past maintenance history may dictate whether it is a candidate for the predictive program. With experience and accurate machinery history records, personnel can effectively direct the program's resources where they will be best utilized. A machine which historically has had very few maintenance problems, requires little attention, and just continues to operate as designed should be considered for only basic monitoring of critical parameters at infrequent intervals. In contrast, a machine which has a history of failure and adversely affects the process should be included in the predictive process to provide the early warning of possible failure and can be repaired during a scheduled shutdown.

Some equipment is supplied with recommendations from the manufacturer describing the parameters to monitor, the frequency of monitoring, and the allowable limits. These manufacturers' recommendations should be incorporated into the predictive program for that piece of equipment. In some cases a phone call or written request to the manufacturer will provide recommendations of parameters to monitor.

Many facilities will find that as they develop their predictive program they do not have the resources available to include all of the equipment, measurement points, parameters, or optimum monitoring frequency to achieve a complete monitoring program. When first establishing a program it is advantageous to start small and grow gradually. This approach provides time to establish the process and make adjustments while the program is more manageable at the smaller size and complexity. Trying to take a program too far, too fast can lead to failure. As every facility is different, every predictive program is unique and must be developed, implemented, and periodically modified as it grows. It is much easier to make these initial modifications while the program is still small and more easily managed.

### Select predictive technologies

Once the equipment to be included in the program is selected, the predictive technologies which are appropriate for each piece of equipment must be determined. This process involves a number of considerations including the following:

Equipment's likely failure mode

Manufacturer's recommendations

Maintenance and failure history

Benefit versus cost of monitoring

Preventive maintenance schedule and activity

After evaluating these areas, the applicable predictive technologies should become apparent. Again, common sense should prevail. If, for example, the temperature of a machine bearing increases and it has been predetermined that no actions will be taken until the vibration indicates action is necessary, perhaps temperature measurements are unjustified.

### Establish monitoring frequency

With the appropriate monitoring technologies identified for each piece of machinery, the next step is to select the proper monitoring frequency. Throughout the initial setup of the program it is important to remember that all aspects of the program can, and should, be modified periodically. Therefore, it is not critical that the optimum mix of technologies or monitoring frequencies be established initially. When selecting the frequency for monitoring equipment the following must be considered:

Criticality of the equipment

Maintenance and failure history

Preventive maintenance schedule

Lead time to failure after fault initiation

Benefit versus cost of monitoring

If resources permit, it is better to monitor more frequently at the start of data collection to establish a reasonable baseline value for each of the parameters monitored on each of the measurement points. Along with selecting the frequency of monitoring, the actual measurement points need to be identified. Of critical importance is the recording of data from precisely the same point each time. This will help to ensure accurate and reliable information from which maintenance decisions can be rendered.

### Select instrumentation

At some point in the setup process the determination must be made as to whether the program will be administered in-house or contracted from a supplier of the service. Factors such as cost, program ownership, and the size and complexity of the facility need to be considered. Some technologies are best applied through the use of a vendor. Oil analysis, for example, is often performed by an established oil lab and only the sampling is performed in-house. For those technologies that will be applied in-house by the facility's personnel, the necessary instrumentation hardware and software must be evaluated and purchased. Many vendors are available to supply both monitoring and diagnostic hardware and software. Some of the same vendors, as well as others, will provide the service under contract.

### Implementation program

Once the program completes the initial setup, implementation begins. The implementation may be accomplished in phases to limit initial costs, provide training time, and to allow for modifications during the learning process. As mentioned earlier, it is often advantageous to start small to give the program the best chance of success.

The initial recorded data, whether it is vibration, temperature, or oil related, will be used to establish baseline levels to which subsequent data can be compared. Ideally, baseline data should be recorded when the equipment is installed and acceptance testing complete. However, reality doesn't always provide that opportunity. Experience has shown that during initial data collection there will be what seems like a large number of parameters which are above alert levels or appear out of the expected range. This is typical and as the program matures and initial problems are addressed, incidents of exceptions will decrease and stabilize. As significant maintenance is performed or equipment replaced new baseline data must be recorded.

### Set alert levels

Alert levels must be established for each parameter measured to provide a flag or trigger for additional action. Initially these levels may be established based on recommendations from the equipment manufacturer, industry standards or guidelines, or the instrument suppliers. Once a representative number of data are recorded and trends of typical operating levels are established, the alert levels can and should be evaluated and adjusted. These modifications will prevent the repeated flagging of equipment which is operating satisfactorily when alert levels are set too low. It will also minimize, for levels set too high, not being alerted to potential problems.

### Measure program effectiveness

As with any program, the predictive program must be monitored for effectiveness. This will provide the opportunity to identify the value of the program, compare it to the other aspects of the integrated maintenance functions, and provide a basis for justifying and measuring expansion of the predictive effort. Programs that add no value to the process will eventually fail. Therefore, it is necessary to document and even publicize the value of the predictive program if it is expected to survive. Knowing how the program compares in value to the other maintenance functions provides insight into how resources should be directed.

Measurement of the program's effectiveness can be accomplished in a number of ways including:

Overall maintenance expenses

Unscheduled repairs

Process downtime

Savings from deferred maintenance

Avoided catastrophic failure

The measurement process may seem unnecessary at first, but it will prove valuable when modifications to the maintenance program are considered and when justification of the predictive program becomes necessary.

### Review and revise program

The predictive maintenance program of any facility should be viewed as a living program. It should be routinely evaluated for its impact and effectiveness. The entire process should be periodically scrutinized to determine whether the appropriate equipment is monitored, the proper parameters and measurement points are included, and whether the proper frequency of monitoring is applied. This will help to ensure the resources are being utilized fully and not being expended on unnecessary monitoring or analysis. After all, the predictive maintenance philosophy embraces performing maintenance only when required without sacrificing reliability. To achieve this, the program must be reviewed and revised periodically.

# NOTES

## NOTES

Chapter

# 5

# Operations and Maintenance Procedures

## Equipment Inventory

Easily the most important aspect of developing a comprehensive operations and maintenance procedures plan, the initial inventory of building systems, equipment, utilities, and controls provides the basis for all work plans, maintenance schedules, and specific tasking and operating procedures. As all preventive and predictive maintenance, corrective maintenance and operating plans emanate from an understanding of the facility equipment and systems needs; this, then, is the first step in the development of comprehensive operating and maintenance procedures.

While extremely important, this aspect of facility management is often neglected or given resources that are not consistent with the level of importance. Many organizations and facility management staffs, whose responsibilities include the development of operating plans and procedures, apply less than adequate professional resources to this effort, resulting in operating plans and procedures that are deficient with respect to the building systems and equipment.

Comprehensive facility management plans require that all of the operating parameters be considered when establishing work plans, priorities, and strategies to affect the long-term utilization of the facility. The equipment inventory should then be considered as part of an overall intake assessment to develop these strategies.

The collection, identification, and planning aspects for facilities management begin with a careful review of all operating systems and equipment, and the needs they are expected to satisfy. The inventory is therefore only one small piece of the intake process. All of the following aspects of facility management need to be considered when performing this function:

1. What are the needs of the facility? More specifically, what are the individual requirements of each functional area within the facility?
2. Do the mechanical, electrical, and control systems meet the needs of the functional areas of operation? This needs to be viewed from both the aggregate and location specific needs. It is insufficient to identify that the ventilation provided by an Air Handler meets the aggregate cfm per square foot requirement of the space being served if specific zone requirements cannot be met with the existing distribution.
3. Does the configuration of mechanical, electrical and control systems meet the needs of the facility in a cost-effective manner? Often, to accommodate specific zone related issues, large generation systems are operated when compartmentalized systems would offer a more effective solution.
4. What is the overall quality of the operations and maintenance practices? What are the current operations and maintenance standards for the building systems? Have the standards been applied uniformly? Are the standards representative of the needs of the equipment? Should the maintenance tasking or frequencies be modified to address specific equipment needs?

5. Do the building systems and equipment require operating engineers and maintenance mechanics specialized training in order to provide the needed operational oversight or maintenance practices?

The issues and questions raised above are not limited to the development of an operating strategy for newly constructed or acquired facilities, rather they serve as functional reminders in any facility operating program. Buildings are collections of dynamic processes that require regular and recurring reviews to ensure the adequacy and appropriateness of intended operating plans. Much in the same respect as preventive maintenance addresses the recurring needs of the systems and equipment in place, a regular assessment of the ability of the installed systems to meet the functional needs of the facility is just as important.

Therefore, the inventory process can be defined as a two-part process. The first aspect deals with the ability of the equipment and systems to function as designed, while the second part deals with the appropriateness of the design itself.

## Conditions Assessment

The conditions assessment process serves to define the needs of the installed systems and equipment. By visiting each piece of equipment, the facility manager establishes a baseline of operating conditions, remaining useful life of the device, and general understanding of the level of maintenance performed to date and whether the device is operating in accordance with the design. This visit is equally important for newly commissioned installations, where the systems and equipment may not have been subjected to the true loads of the facility during the commissioning process. A recommended frequency of this type of assessment is annual. The focus of the assessment should include:

1. Documenting the working order of the equipment and systems.

2. Determining if the frequency of maintenance provided is appropriate. Should the frequency of maintenance visits increase or decrease?
3. Determining if the level of maintenance being performed is thorough enough to maintain the equipment in sound operating condition. Should additional tasks or tests be added to the current preventive and predictive maintenance checklist?
4. What is the quality of the maintenance performed? Is it lacking or acceptable?
5. Are there any operating or maintenance deficiencies that need to be addressed?
6. How critical are the deficiencies? Should they be performed immediately or within a reasonable time frame?

The following checklist serves as a guideline for the performance of the conditions assessment. Items to be visited should include mechanical, electrical, lighting, plumbing, utility, controls and building automation, fire/life safety, doors, roofing, and structural components.

Create a baseline of the current conditions by:

- Documenting the working order and condition of all equipment and systems
- Documenting items missing, such as belt guards, fan shrouds, or electrical box covers
- Documenting any and all deficiencies noted and generating a deficiency report
- Prioritizing the deficiency list
- Identifying which deficiencies can be repaired during the course of routine preventive maintenance tasks versus those that require additional labor or specific expertise
- Developing a performance cost estimate for the identified deficiencies
- Documenting the current quality level of operations and maintenance

- Reconciling the equipment inventory to the CMMS modules
- Identifying any devices which require increased or decreased frequencies of maintenance
- Updating the CMMS modules to reflect any changes needed in the maintenance practices
- Documenting the estimated remaining useful life of the equipment in the CMMS
- Insuring that the CMMS database matches the nomenclature of the systems and equipment installed
- Identifying suspect operating conditions, or systems, which may warrant professional inspection

## Design Assessment

This aspect of the conditions assessment is important as many facilities were designed and constructed in an effort to accommodate multiple tenant usage patterns. Particularly true of facilities constructed as speculative office facilities, the original engineering design was developed without any specific information on the tenant needs. Beginning with a cold, dark shell (no HVAC or reflective ceiling plan), many assumptions regarding the aggregate electrical and air-conditioning loads were incorporated during the design process. Facilities constructed in this manner up to the late 1970s and early 1980s often were based on anticipated loads that are no longer valid assumptions in today's electronic offices. Conventional approaches used in estimating the occupant and appliance loads were one person for every 200 $ft^2$ and 0.5 W of receptacle power per square foot, respectively. The advent of compartmentalized office furniture systems in open floor plans with personal computers at each workstation have easily rendered those assumptions invalid.

Even newer facilities may suffer from an inability to meet the current intent of ventilation codes and standards adopted to achieve enhanced indoor air quality. Where only a few years ago, energy conservation was given more attention than the concern for occupant health, ventilation rates in most facilities constructed throughout the 1980s are deficient with respect to the current trend in the development of ASHRAE standards.

The concern over occupant health is not likely to wane in the foreseeable future, so the challenge facing facility managers and designers today is to meet these increased ventilation requirements without incurring substantial operating cost increases. An indoor air quality plan should be developed and implemented to ensure that the facility has healthy buildings. Systems to be evaluated include those identified in the conditions assessment—only the focus of the evaluation is related to code compliance, occupant safety, and energy performance.

In performing these assessments, several key questions should be addressed:

1. Have any areas of the facility undergone substantial changes in operating hours, deployment of personnel, or electrical loading?

2. Have the building codes been modified since the construction of the facility?

3. What changes are necessary to bring the existing systems and equipment into compliance with the updated standards and codes?

4. Is the existing control system capable of operating within the intent of the updated codes and standards?

The result of this evaluation should form the basis of a capital improvement plan, designed to modernize the facility, improve indoor air quality, and promote occupant safety and health while mitigating unnecessary operating or energy expense. These objectives are not mutually exclusive where the modernization includes the necessary controls and automation to insure compliance with the program objectives. Some modernization measures may be attainable without expensive modifications to the HVAC systems and equipment, and can be implemented through changes in the base operating plan of the facility.

## Operational Requirements

Forming the basis of facility operating procedures, the *Building Operations Plan* (*BOP*) outlines standards and guidelines for operating buildings' systems and equipment such that all temperature, lighting, and other environmental conditions required by individual zones are provided during normal building operating hours. The specified guidelines establish the desired conditions based on time of day, weekdays, weekends, and holidays, and reflect the desires of tenants, departmental objectives, or occupant agency directives. The BOP provides guidance for developing maintenance and operational staffing levels to facilitate the needs of the tenants or occupants within the framework of the mechanical systems and equipment installed at the facility. In office buildings, it also serves to define the separation between building standard operations, and those additional requirements that specific tenants may require.

Key to BOP development is achieving understanding of the needs of the tenants, the resultant requirements of the mechanical and electrical systems and equipment, and the level of staff needed to insure the safe and effective operation of the building systems and equipment. As there are a wide variety of building designs and operating conditions, the following plan is just a sample of the elements that are routinely addressed in the BOP. Conditions affecting the actual plan include weather, equipment condition (which is strongly influenced by prior maintenance activities, equipment installations, operations, etc.), tenant-use patterns, and special requirements, in addition to a host of other variables. Therefore, the BOP is a dynamic plan that must be able to respond to many varying conditions. Facility personnel constantly adjust and modify the plan, as needed, to respond to severe weather patterns, extended operational needs by tenants and changes in the performance characteristics of the buildings systems and equipment. Items regularly addressed in the operational plan include:

Equipment start-up and shutdown procedures

Building temperatures (both general and special areas)

Operational checks

Energy conservation requirements and objectives

Use of fresh air and economizers

Air filtration

Potable hot and cold water

Lighting levels

Relamping

Controls and automation systems

In general, the BOP addresses the intent of the facility manager's policies and procedures to ensure tenant comfort and safety while minimizing operating expense. Further, it serves to define acceptable practice for the scheduling of maintenance activities including operational tests of large energy consuming devices. Examples of policy and procedural objectives include:

- Systems and equipment will be maintained in the automatic mode to the maximum extent practical.
- Routine procedures will guide start-ups and shutdowns.
- Equipment will not be cycled at frequencies that lead to premature failure.
- When duplicate systems exist, equipment will be rotated to equalize usage. Another school of thought on this subject is to run one of the two devices much more than the other, so that both will not reach the end of their useful lives at the same time.
- Utility meters will be read on the same date as they are read by the power company.
- Inventories will be maintained for all fuels received, stored, and used.

- Inventories will be maintained for all refrigerants received, stored, and used.
- Preventive and predictive maintenance and testing of large electrical devices will be scheduled at a time when the operation of the device will not impact the electrical demand peak.

## Equipment Start-up and Shutdown

Daily facility start-up procedures will vary according to external environmental conditions and average building space temperatures. In general, the air-conditioning and heating equipment will be started (and shut down) at the same time each day. Start-up times should be coordinated with the arrival of the start-up operating engineer, whether the equipment is started automatically via time-clock or energy management control system. Prudent operating practice includes the presence of the operating engineer during primary machinery operations.

The following tasks should be undertaken at start-up:

- Upon arrival, the start-up operating engineer should assess external environmental conditions (weather, light, etc.), and any other building-use factors to ensure that the controls (Energy Management Control System or time clocks) have started or will start the mechanical equipment in order to have the building within prescribed environmental limits by the start of normal building operating hours.
- Observe the mechanical system start-up. Inspect the systems as they become operational, noting excessive vibration, and equipment operating temperatures and pressures. Verify chiller start-up purge operations and transition.
- Perform a watch tour of the systems after start-up to insure the proper operation of the equipment.
- Any major equipment or systems, including elevators and security and fire alarm systems, found to be inoperable need to be addressed immediately by departmental personnel or contractor labor. The intent of the start-up process

should be to provide a reasonable amount of time to assess inoperable equipment and take corrective action before tenant comfort or safety is compromised.

Tables 5.1 and 5.2 provide a brief description of start-up measures taken, during each season of the year, to ensure that interior building temperatures and environmental conditions adhere to tenant requirements. Equipment start times and settings will be governed by the following conditions:

- Outside air temperature
- Perimeter and interior building space temperatures
- Weather forecast for the day
- Current weather conditions (i.e., precipitation, sun load, humidity, wind conditions
- Time elapsed since last operation (and/or weather change)

Shutdown procedures (which follow below) are designed to ensure that the systems and equipment are properly secured and that the desired mode of operation for the ensuing start-up functions can be achieved (see Tables 5.3 and 5.4).

- Before the shutdown operating engineer leaves for the day, he will review the weather forecast for the evening and following day, indexing the equipment to the mode of operation that will best meet the needs of the facility the following operating day.
- Equipment should be indexed for automatic operation through the EMCS or time clocks as to when to initiate shut down procedures and when to schedule the subsequent start-up.

Note that these procedures ignore the Energy Management Control program function of optimal start and optimal stop. These functions can be automatically initiated through most Energy Management Control Systems. Care should be taken with respect to this operation as Energy Management Control programs can only make decisions based on information they possess at any point in time. They

**TABLE 5.1 Typical Summer Start-Up Guidelines**

| | Outside air temperature | Perimeter temperature | Start-up time |
|---|---|---|---|
| 1 | 60° or below | Below 76° | Fans as needed after 7:00 A.M. for free cooling, chiller as needed after 7:00 A.M. |
| 2 | 60° or below | Above 76° | Fans at 6:30 A.M. for free cooling, chiller as needed after 6:30 A.M. |
| 3 | 60° to 70° | Below 76° | Fans at 6:30 A.M. for free cooling, chiller as needed after 6:30 A.M. |
| 4 | 60° to 70° | Above 76° | Fans at 6:30 A.M. for free cooling, chiller as needed after 6:30 A.M. |
| 5 | 70° to 80° | Below 76° | Fans 6:00 A.M., chiller 6:00 A.M. |
| 6 | 70° to 80° | Above 76° | Fans 6:00 A.M., chiller 6:00 A.M. |
| 7 | Above 80° | — | Fans 6:00 A.M., chiller 6:00 A.M. |

NOTES:

1. During prolonged shutdown (weekends, holidays) it may be necessary to advance the start-up schedule by several hours.
2. During extreme heat wave conditions it may be necessary to operate for prolonged periods.
3. These are only typical guidelines, and need to be adjusted to the need of the facility.

**TABLE 5.2 Typical Winter Start-Up Guidelines**

| | Outside air temperature | Perimeter temperature | Start-up time |
|---|---|---|---|
| 1 | 50° and above | Below 72° | Fans and heating equipment at 7:00 A.M. |
| 2 | 32° to 50° | 68° to 72° | Fans and heating equipment at 6:30 A.M. |
| 3 | 32° to 50° | Below 68° | Fans and heating equipment at 5:00 A.M. |
| 4 | 27° to 32° | Below 68° | Fans and heating equipment at 4:30 A.M. |
| 5 | 27° to 32° | 68° to 72° | Fans and heating equipment at 5:00 A.M. |
| 6 | 27° and below | Below 68° | Fans and heating equipment at 4:30 A.M. |
| 7 | 27° and below | 68° to 72° | Fans and heating equipment at 5:30 A.M. |

NOTES:

1. During prolonged shutdown (weekends, holidays) it may be necessary to advance the start-up schedule by several hours.
2. During extreme cold weather conditions (extreme cold, high winds, severe wind chill, heavy snow, etc.), it may be necessary to operate for prolonged periods or around the clock.
3. These are only typical guidelines, and need to be adjusted to the needs of the facility.

**TABLE 5.3 Typical Summer Shutdown Guidelines**

| | Outside air temperature | Perimeter temperature | Shutdown time |
|---|---|---|---|
| 1 | 60° or below | 76° | All equipment 5:05 P.M. |
| 2 | 60° to 70° | 76° | All equipment 5:05 P.M. |
| 3 | 70° to 80° | 76° | All equipment 5:05 P.M. |
| 4 | Above 80° | 76° | All equipment 5:05 P.M. |

**TABLE 5.4 Typical Winter Shutdown Guidelines**

| | Outside air temperature | Perimeter temperature | Shutdown time |
|---|---|---|---|
| 1 | 50° or above | 70° | All equipment 5:05 P.M. |
| 2 | 32° to 50° | 70° | All equipment 5:05 P.M. |
| 3 | 27° to 32° | 70° | All equipment 5:05 P.M. |
| 4 | 27° and below | 70° | All equipment except freeze protection equipment and heating pumps, which run all night—5:05 P.M. |

cannot obtain forecasts and adjust program objectives on data that have not been received. This can result in increased costs of operation by failing to enact precooling modes that serve to mitigate anticipated demand peaks. As most utility structures penalize the demand aspect of electric service more than the consumption element, it is routinely more cost effective to start equipment early and avoid demand peaks than to delay the start-up when demand will become an issue later in the day.

Severe weather situations may dictate continuous operation of the systems for freeze protection.

## Operational Checks

Operating watch tours should be constructed so as to confirm the operations of the mechanical systems and equipment are in accordance with the BOP specifications. Not limited to conformance with plan objectives, the operating watch tour is a proactive assessment of the ability of the systems and equipment to meet the needs of the facility and insure the useful life of the equipment. These tours should:

1. Identify the operating parameters of the major building systems
2. Provide feedback as to the ability of the systems to meet the needs of the facility
3. Identify areas where the systems and/or equipment are not being operated or maintained in accordance with the established objectives
4. Identify areas where conditions warrant modifications to the operating parameters
5. Identify excessive vibration or motor operating temperatures that will inhibit the ability of the systems to maintain reliable operations
6. Identify opportunities to enhance the energy conservation of the facility
7. Identify equipment and system deficiencies that will result in unscheduled repairs and/or failures that jeopardize the life of the equipment
8. Identify lighting systems and equipment that fail to meet to meet the needs of the usage areas

## Lighting

Lighting systems are designed to meet the aggregate requirements of the tenant spaces. In general, the standards for illumination vary with the intended usage of the space. The following illumination levels represent normal light levels required to facilitate office and shops usage and provide tenant safety.

Illumination levels during tenant normal work hours are:

| | |
|---|---|
| Public areas within buildings | 10 footcandles |
| Normal workstations | 50 footcandles |
| General work areas | 50 footcandles |
| Storage areas | 10 footcandles |

These lighting systems standards should not be increased, or decreased, without consideration given to all requirements within the areas being modified. In case of insufficient light levels, it is far more productive to combat the specific need with task lighting than to increase the aggregate lighting levels of an entire area to accommodate a specific function. The prescribed lighting level footcandle readings should be randomly spot-checked during quality control inspection tours and/or operating watch tours. Degradation in the illumination levels can usually be linked with a need for cleaning of the fixtures or the aging of the lamps. Periodic checks of the illumination levels will serve to identify the rate at which the lighting levels are degrading.

### Relamping

When areas of the facility indicate lumen degradation to a level that hinders the safe and secure usage of the space, group cleaning or relamping should be scheduled. As the major expense associated with either activity is labor, relamping prior to the calculated end of the useful life of the lamp should not be a significant deterrent.

## Energy Conservation

Temperature controls should be set to maintain an indoor environment consistent with general industry standards and practice during normal facility work hours. With the exception of specific areas that may require special environmental conditions, temperature controls should be maintained between 72°F (22.2°C) and 76°F (24.4°C). This temperature control range is understood to be the generally acceptable temperatures established for human comfort. Operating outside of this range under the auspice of energy conservation usually results in excessive temperature complaints by the tenants. Conservation should be an attempt to maximize the effectiveness with which desired conditions are met, not an effort to force environmental conditions on the tenants that are undesirable.

In maximizing this effectiveness, special consideration should be given to the use of economizers or free coolers to meet the needs of the facility. In general, the operating plan should emphasize the use of air and water side economizers to the maximum extent practical. Mechanical cooling and heating operations should be limited to the occasions when these economizer systems cannot maintain the desired tenant conditions.

Building operating policy should include the following objectives:

1. During moderate weather, use outside air, mechanical economizers, and/or other energy-saving equipment installed, to the maximum extent possible.
2. The use of outside air, equipment, or similar strategies should be based on outside temperature and humidity levels.
3. Building ventilation should be provided to the maximum extent allowed by the configuration and design of the installed mechanical equipment.
4. Fresh air should be adequately filtered at all times by using only those air filters capable of at least 50 percent particulate removal. Filters should be changed frequently, in accordance with best industry standards and practices.
5. In the case of conflict between what is installed in the building and what the building design specified, the more stringent air requirements should apply.
6. Ventilation standards are defined and specified in the EPA's Clean Air Act and OSHA's Indoor Air Quality Standards; Quality Standards for Design and Construction Handbook PBS P 3430.1, Appendix 5-R; the Energy and Water Handbook PBS P 5800, Chapter 8-7e; and in the American Society of Heating, Refrigerating, and Air Conditioning Engineers Standard 62-1981, Ventilation for Acceptable Indoor Air Quality; and CFR Title 40, Part 141 PCB procedures.

It is important to note that ASHRAE standard 62 is currently being revised to create a more stringent requirement for minimum fresh air levels in facilities. This is in response to the growing number of indoor air quality complaints across the country.

## Demand Control Ventilation

With the focus on increased ventilation rates, many facility managers are turning to *demand control ventilation* (*DCV*) as a means of identifying and controlling the level of fresh air provided to a facility. DCV is a method of modulating fresh air intake to a facility in response to the level of carbon dioxide ($CO_2$) present in the building system return air. In theory, the DCV application serves to increase fresh air intake volumes when $CO_2$ levels rise above a setpoint (nominally 500–600 parts per million). Concentrations of $CO_2$ above these values is construed as insufficient oxygen for human comfort and safety. Clearly, sampling $CO_2$ can provide a measure of the amount of fresh air that is being utilized for breathing, but does not indicate other needs of the fresh air. Expulsion of outgasses commonly found in facilities is one of the most important elements of indoor air quality. Using only $CO_2$ as the barometer for controlling ventilation rates could result in failing to address one of the prime candidates for IAQ complaints.

While growing in popularity, this application is not a sure cure to the issue of ventilation standard compliance. The ASHRAE standard 62 relates to specific ventilation rates required for specific functions within a facility. These rates are defined as cubic feet per minute (cfm) per person or per square foot of area, varying with the type of occupancy and usage. Using aggregate sampling at common building exhaust or return air systems could indicate an aggregate compliance with the provisions of the standard, while specific areas within the facility could be under served. Another major challenge to the utilization of this technology rests with the accuracy of the sensors themselves. Commercial grade apparatus is commonly available with an accuracy of

5–10% of the range of the device, with specified minimum values of potential offset (e.g., ±5% or 75 ppm, whichever is greater). When considering the values being measured, a 5% inaccuracy for a reading of 600 ppm (with the stipulated minimum error of 75 ppm) is actually an error of 12.5%. While this may represent a reasonable degree of measurement, using this value as a control point could lead to inappropriate control actions or the creation of a false sense of security. If the application of this technology were contested as the cause of IAQ related illnesses, it would be difficult to defend this range of accuracy as a means of addressing such a prominent tenant health issue.

## Facility Management Tools

Energy management control systems (EMCS) can be utilized as a valuable resource in performing facilities management. While the systems cannot hear unusual noises or see excessive vibration, they can be utilized to augment and enhance the operational effort. With facility staffing spread thin, the EMCS can be used as a strategic tool to assess the operations and maintenance requirements of facilities. It is invaluable in providing the necessary information to identify problematic areas before they become critical failures.

Conventional thinking applied to EMCS is focused on the ability of the connected points to be controlled, thereby controlling expenses. This strategy only addresses the utility cost portion of the operating expenses. While important, the most disruptive element in facilities management is the unplanned expense.

Utility expenses can be predicted with reasonable accuracy accounting for variances in weather, occupancy, production, rate schedules, and usage changes. Using historical data, these variances can be projected to establish a reasonable expectation of utility consumption and costs. Unplanned repairs, however, are not so easily predicted. They are dependent on the following issues:

1. The quality of maintenance being performed

2. Quality of equipment being used
3. Effectiveness of the system design
4. Level of staffing (adequate to maintain the systems)
5. Complexity of the systems and equipment
6. Ability of the operating personnel to diagnose the causes of the problems

Facility management systems can be utilized to provide the needed information to assess the conditions noted above. By extending the telemetry to include nonenergy conservation monitoring and control points, the EMCS can become a management information system for system reliability and operations oversight.

Specifically, poor maintenance practices on mechanical systems and equipment can be determined by monitoring vibration, motor temperatures, oil temperatures, or the effectiveness with which the equipment is performing its required functions. BTU meters measuring the delivered chilled water and hot water can be compared with the electricity, oil, or natural gas utilized to create the commodity resulting in the development of an overall effectiveness of generation. As heat transfer surfaces begin to foul, the effectiveness of the system will diminish. This is indicative of poor maintenance to the heat transfer surfaces. Corrective measures needed to overcome the poor transfer rates would include coil cleaning, eddy current tests and tube cleaning and heat exchanger tube cleaning. Performing maintenance based on the degradation in the effectiveness would prevent emergency repairs needed when the quality of heat transfer degrades to a critical point, where systems or equipment could fail.

Similar evaluations of the effectiveness of HVAC systems to meet the needs of the facility may indicate design issues that are not in accord with the needs of the facility. The design condition assessment can be augmented dramatically using the EMCS as the collection point for system operating profiles. A survey by consulting engineering personnel typically involves a walkthrough of various systems with small samples of data taken form the operating systems during the brief visits to the machinery. The EMCS clearly can provide historical measured values for the same piece of machinery on an extended basis. Trend logs and point histories then become the basis of a perpetual survey of operating issues.

Information unto itself is useless unless the operating personnel are trained in the analysis of the data and the impact it has on the facility requirements. Careful consideration should then be given to the amount of data to be collected, the frequency of the collection, and the results or level of analysis desired. As with any computer system, there is a finite amount of processing power resident in any EMCS. Extending trend logs and point histories to every connected point may diminish the ability of the EMCS to perform other functions, such as operating programs, control loops, and alarm notification functions.

In using EMCS technologies to mitigate the unplanned maintenance and repair expenses, it is therefore critical that the EMCS itself become the most reliable operating system within the facility. This is easily accomplished by applying the same level of maintenance planning, scheduling, and tracking as is applied to the systems the EMCS controls and monitors. Most facility maintenance management programs can easily accommodate the EMCS monitoring and control points as part of the equipment inventory. The level of maintenance required to ensure the proper operation of the EMCS is fairly consistent across the multitude of vendors and manufacturers. In the same fashion that mechanical systems are defined as generation, distribution, or terminal use components within the overall HVAC system, the EMCS architecture can be defined as:

1. *Workstation or head-end devices.* Comprising the computer, operating systems, man–machine interface, display terminal, remote interrogation devices (modems), printers, and communication devices to poll the data collection/gathering panels. See Fig. 5.1.

2. *Data-gathering panels.* Whether for data collection or command processing, this device represents the primary component in the EMCS architecture. Typically a standalone capable device, it can measure and record values, execute program instructions, and command field devices without the presence of the operator workstation. See Fig. 5.2.
3. *Field devices.* These are the relays (start/stop or two position status), variable condition sensors (temperature, humidity, pressure, etc.), and variable output devices for positioning valves dampers and other control devices.

## Energy Management Control System (EMCS) Operations

EMCS is a computer-based energy and temperature control system, featuring microprocessor-based remote field panels capable of standalone operation, that is used to administer building operations. These systems typically have modular architecture permitting expansion by adding computer memory, application software, peripheral equipment, and field hardware. Whether identified as an EMCS, EMS (Energy Management System), BAS (Building Automation System), or FMS (Facility Management System) these systems consist of the following components:

- Building environmental monitoring and control system, subsystems, and associated equipment
- Central control center console system, subsystems, and associated equipment
- Software programs (internal)
- Energy management control system, subsystem, and associated equipment

While operation of the FMS generally falls to the facility department's operating engineers, maintenance and repair services for these critical building systems are subcontracted—usually to the manufacturer of the system. As these

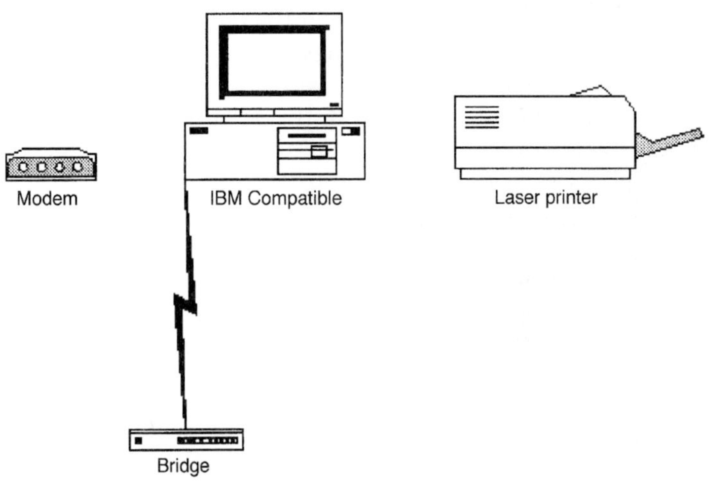


**Figure 5.1** Workstation architecture.

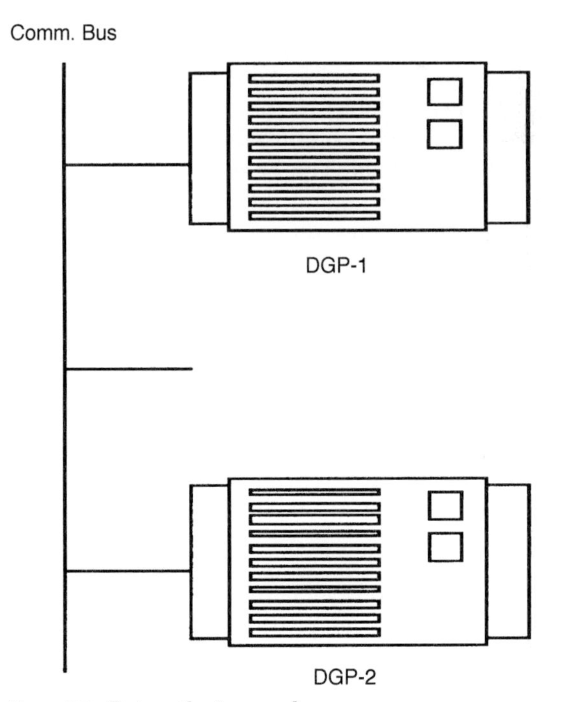


**Figure 5.2** Data-gathering panels.

systems perform a vital service in the operations and maintenance of facilities, they are deserving of a level of attentiveness consistent with the mission.

Understanding the pieces of the system and how they relate to the successful implementation of the building operating plan is essential to managing this valuable resource. In developing this aspect of the building operations plan, it is important to identify the tasks requisite to properly maintain the FMS and the craft/trades best suited to accomplish these functions. Far too often, facility operators rely exclusively on the installing manufacturer for all aspects of system maintenance. This practice translates into higher maintenance costs and fosters an extremely dependent relationship between the owner and operator with the system manufacturer.

FMS manufacturers are well versed in the requirements of their individual systems, products, and programs. They should not be construed as being experts in the application of HVAC control technologies, energy strategies, or control engineering. They are product experts and should be relied on to maintain the portions of the product that departmental operating engineers cannot.

As noted above, the FMS is combination of computer technology and electronic controls. The head-end computers are generally off-the-shelf components with custom software applications resident to poll the remote panels, follow operating sequence programs and effect monitoring and control commands. The remote panels are essentially data collectors and microprocessor-based receiver controllers that emulate the same basic logic of conventional temperature controls. The primary benefits between these systems and the conventional control products are:

1. The ability to embed many receiver controller functions within one microprocessor
2. A higher degree of accuracy without constant calibration
3. The ability to alter or modify the operational response of many control loops from one location
4. The ability to monitor the operating parameters of an entire facility from one location

5. The ability to coordinate the operations of many devices in a manner that promotes energy conservation, while maintaining required comfort conditions

These benefits are mostly related to operational enhancements in the effort to mitigate energy usage and/or demand. When integrated as part of the building operations plan, they are utilized to perform the following routine functions:

- Start and stop the building HVAC equipment according to predetermined schedules established in the building operations plan
- Monitor and log any abnormal operational readings of the building systems and equipment
- Assess the operating efficiencies of the mechanical systems by developing operating profiles of the systems over varying load conditions

While these systems are vital to the effective operation of the facility, an emergency plan of operation must be developed for those instances when the FMS is out of service or compromised.

## Emergency and Extended FMS Operations

As the FMS is a vital tool in the performance of facility management, its reliable operation is critical to the successful facility manager. As such, the ability of the FMS to perform its functions cannot be compromised. The start-up operating engineer arriving on site in the morning must know that the FMS is available to perform the requisite operating functions in order to allow the operating engineer to perform watch tours and other engineering functions. Finding the FMS in an inoperable condition at the appointed start-up time would create a tremendous burden on the operating engineer to manually start the needed plant equipment.

To ensure the ability of the FMS to perform its functions, it is prudent to have the system operating status monitored continually. Any number of potential solutions exist to monitor the operating condition of FMS equipment, from autodialers to remote monitoring services. In either event, the intent of the service is to notify operating engineers personnel that the FMS is incapable of performing automated start-up functions.

Whether this monitoring is performed by on-site personnel or remotely, the facility manager must provide emergency operations support in response to facility needs. The criticality of the facilities operations will dictate the level of attentiveness required to ensure the appropriate response. In most instances, it is understood that an emergency is defined as a situation which constitutes an immediate danger to personnel, a threat to damage property, or a threat to disrupt occupant operations. Examples include outages in utility systems, clogged drains, broken water pipes, gas leaks, and electrical distribution problems. Incidents such as these require an immediate response to protect the property. Overall responsiveness is then dependent on the extent that the FMS can automatically notify facility management personnel. Key elements of this responsiveness are outlined below:

- 24 hours per day, 365 days per year emergency call service support in response to emergency conditions. This would include operating the facility during emergency conditions via the FMS. It follows that service calls to restore full FMS operations would be treated as an emergency.
- The responding technician or craft/trade must be knowledgeable in the operation of the FMS. Using the FMS as the data collection and assessment tool, the responding tradesman would:
  1. Detect, locate, analyze, report on, and correct, as required, abnormal conditions in the mechanical systems. Typical cases might include freezing temperatures, high winds, snow, heavy rains, and similar adverse weather conditions.
  2. Correct all alarm situations, including emergency shutdown of the facility equipment (neutralizing alarms as necessary prior to correction).

3. Monitor the automatic restart of the FMS computer and facility equipment after a power failure or any other malfunction. During the emergency situation, provide manual operations, as required, to maintain facility support functions.

Apart from the specific functions noted above, the FMS also provides other forms of support during emergencies. The FMS can be utilized to create an audit or review of the incident. Utilizing trends, point histories, and program analysis, the FMS can provide information on the operating conditions of the equipment and systems prior to the failure. This type of analysis is often very valuable in determining corrective measures to preclude the recurrence of the failure. Used properly and to the full capability of the system, the FMS is an extremely valuable tool in the performance of operations oversight, quality control, and facilities management.

## FMS Service Program

The successful facility manager will maintain all EMCS hardware, software, and associated components in accordance with the manufacturer's specifications and sound industry practice. Proper system maintenance will include preventive maintenance (PM), corrective maintenance, software support services, and analysis of declining conditions that warrant further study. All of these elements together represent the requisite attention needed to ensure the successful implementation of the FMS.

Understanding the importance of proper FMS maintenance is demonstrated by the life cycle of the energy management systems and equipment. In use every day as a critical tool in building operations, the FMS service program should include the following features:

1. *Daily verification checks on all connected points.* Every connected point should be regularly checked for accuracy against portable measurement standards to ensure the continued reliability in the sensed values. This is critical to the successful performance of operating programs that base the automatic calculation of setpoints or strategies on the

measured values from the field points. Points determined to be "out-of-tolerance" need to be recalibrated to return the level of credibility requisite of the control strategies.

2. *A highly disciplined corrective maintenance program.* Central computers, hard disks, printers, and video display terminals should receive monthly inspections to ensure continued reliability. Connected points should be reviewed daily by FMS operating personnel to determine if they are responsive and provide reliable feedback. All points determined to be "suspect" should be tested by O&M personnel in conjunction with the FMS operator. Points that are determined to need corrective action should be either immediately recalibrated or noted on a service/repairs list. This list should be maintained in a database, or computerized maintenance management system, to ensure adequate oversight. Note that the onus is placed on O&M personnel to make the initial determination of point trouble to ensure that only truly "bad points" are placed on the point service list. This practice aids in avoiding unnecessary, costly repairs by subcontractor personnel to issues that could be easily handled by site personnel.

3. *Diligent FMS Preventive Maintenance (PM) services.* PM work to be provided should include equipment inspections to ensure proper operation. This includes a schedule of routine maintenance (annual work plan) to be applied consistently for all FMS devices. As with mechanical and electrical systems, PM work orders for each task, detailing the work to be performed, required technician skill levels, and any special tools, instruments, and diagnostic programs needed, should be generated for all data collection and transmission gear.

4. *Service repair records.* Dispatched service calls for repairs, calibrations, or parts replacements should require a completed work order for each task completed, reporting on the work accomplished, labor hours, and materials used. This is valuable in assessing recurring issues with specific components of the FMS.

5. *System diagnostics.* These are essential in detecting early signs of deteriorating performance and serve to pre-

dict potential system failures. Potential problem areas should be identified and corrective actions taken, using the most advanced tools, instruments, and diagnostic software.

Much of the tasking elements identified above can be accomplished using facility operations and maintenance personnel. Conventional FMS service plans tend to require the above functions from the manufacturer's personnel. This practice is very expensive and does not relieve the departmental personnel from assisting in the process. The facility management operations engineering personnel, tasked with the proper operation of the mechanical systems and equipment, are in the best position to identify problems and the corrective actions needed. Subcontractor personnel will not be as familiar with the nuances of the facility, equipment, or systems—information critical to the proper analysis of system-related issues.

## FMS Operating Personnel Requirements

This area of facilities management often gets overlooked when developing a comprehensive operating plan. Many facility managers will elect to have system technicians, knowledgeable in the FMS products and programs perform this function. This is akin to having plumbers perform electrical work or elevator technicians cleaning cooling towers. The FMS is no more than a tool to be utilized in the operations of facilities. The commands to be initiated are no different from time clocks or manual start stations. The only logical craft to operate facility management systems is operations and maintenance personnel trained and licensed to operate machinery. As such, the licensing provisions requisite of the operating engineers is the same standard that needs to be applied to the FMS operating personnel. This standard applies whether the personnel are resident in the facility to be operated or remotely located.

Often, facility systems and equipment are monitored by personnel that have no training, education, or experience in the systems and equipment that they are monitoring or controlling. While the practice of remote monitoring and control can lead to significant operating cost savings, the methodology employed in selecting these service providers should be no different than a hiring decision for departmental operating engineering personnel. Some jurisdictions in the country have sought to develop standards for the remote operations of facilities. What follows is a guideline in developing those standards, reflecting the licensing provisions for the Washington, D.C., metropolitan area.

### Remote control operating requirements

In establishing licensing requirements for remote monitoring and control facilities, each operation's aspects must be viewed independently to determine the proper methods, staffing levels, training, and communications to protect the property and ensure safe operation. These issues must be evaluated at both the controlling facility and the controlled facility. It is the intent of these standards to define the manner in which remote control and monitoring may be utilized to satisfy the provisions of DC Code Chapter 4, Section 401-Duties of Licensed Operators. The provisions of DC Code Chapter 4, Section 400 shall apply to the operating engineers utilized in the performance of remote control services.

**Definitions.** The distinction between monitoring and control facilities lies with the fundamental ability to change the operating state of a piece of equipment. Whether an adjustment to a variable position or a change of state (on to off, etc.), the ability to render such an action is the definition of a remote control facility. Monitoring locations, limited to such notices as alarm annunciation and acknowledgment of remotely activated conditions, are not to be considered as remote control facilities.

The term "Licensed Operating Engineer" used herein is defined as a District of Columbia Third-Class Steam Operating license (at a minimum). *Note:* This is the area standard for operating engineers in low-pressure steam plants.

The term "responsible" refers to the licensed operating engineer tasked with the operation and oversight of the remote control functions of a facility. This functional responsibility includes actions undertaken by the operating engineer together with those actions initiated through an automated routine by the remote control system.

"Status" as defined herein refers to the proof of operation of a given device. It is a positive indication that the work the device is designed to perform is being produced. Indicators that fail to positively identify the device as an operating and delivering utility resource do not meet the qualifications as a positive status.

"Certified remote control facility engineer" is defined as an applicant who has been certified to be in compliance with the provisions of the standard.

### Remote control facility provider

To be a certified remote control facility provider, the service provider needs to conform to the following minimum provisions. Conformance includes an initial inspection of the facility, with recertification inspections performed on a recurring basis.

### Staffing issues

**Licensing requirements for operators.** The standard for remote control operating personnel affecting command and/or control initiatives for a property in the District of Columbia is the D.C. Third-Class Steam Engineers license. This measure of the applicant's overall training, experience, and commitment to the trade denotes the minimum licensing of remote operators. Any and all control actions (variable- or two-position) shall be initiated and verified by a licensed operating engineer. The use of personnel that do not meet this requirement renders the operating entity in violation of this standard.

**Staffing levels.** A minimum of at least one licensed operating engineer must be present in the remote control facility at any time that remote control services are being performed.

This requirement holds for all remote initiated commands and control actions, whether initiated by the operator or the automated system in use. Command and control actions undertaken by the automated system are deemed to have been made by the licensed operating engineer with responsible charge of the operation of the particular facility.

**Training and certification.** The operating engineer with responsible charge of the remote control functions shall maintain a valid third-class steam engineers license. Any retesting, recertification, or other requirements which become part of the minimum requirements of new applicants for similar licenses must be met by the responsible charge engineer. In addition to the license, the operating engineer with responsible charge must possess a universal CFC license as defined by the Environmental Protection Agency. *It is the intent of this standard that any training or certification required of field operating personnel to maintain a valid local engineering license shall also be required of the responsible charge engineer.*

**Supervision.** The minimum licensing provisions for the operating engineer will extend to the line supervisors of the remote control functions. A District of Columbia Third-Class Steam Engineers license is required of the line supervisors.

### Systems and communications

**Supervised communication formats.** Communications systems utilized in the performance of remote control services shall be supervised. Supervised communication formats include any and all telecommunication services where the availability of the communications path is supervised by the remote control and monitoring system. Examples of these services include: dedicated lease lines, ISDN, and dedicated fiber-optic networks. Polling systems, whereby the controlled facility only "reports by exception," shall not be considered supervised unless automatic testing of the availability of the communication path is included with the service. Time-based testing of the availability of the communications service shall not be less than four times daily. Evidence of the automatic polling shall be provided to the reviewing entity in the initial certification inspection together with recertification activities.

**Alarm reporting.** Systems which only annunciate alarms in "report by exception" formats shall not be considered as compliant with these standards and will not be certified as remote control facilities. Nothing contained in these standards precludes the ability of a "report by exception" system to be utilized as a monitoring facility only. Systems of this type do not comport with the requirements of remote control operations and shall not be certified as remote control facilities.

#### Contingent operations

**Loss of communications or control.** As remote control services are being provided in lieu of staffed operations, any loss of communications or control by the remote control facility that cannot be restored within 4 hours shall warrant a return to performance of the operating requirements via direct means.

**Electronic logs.** The remote control facility shall prepare operating logs of system parameters and maintain these records for inspection by the reviewing entity.

#### Continuity of service

**Redundancy.** In the event of a loss of control or communications, the remote control facility must allow for continuity of services through the provision of redundant operations. Redundancy must be in the form of an alternate automatic remote controller or the return to local control.

**Emergency services.** The remote control facility must provide for the continuation of operating services in the event of disruption in the normal supply of building services.

**Power.** Sufficient emergency power or an uninterruptible power supply must be provided to maintain monitoring and control facilities in the event of a loss of power to the remote control facility. Emergency power/UPS shall be capable of withstanding a loss of primary utility power for a period of 8 hours.

**Lighting.** In addition to the emergency power requirements of the controlling systems, the provision of emergency services shall extend to the lighting requirements of the remote control facility.

**Air conditioning.** Air-conditioning and ventilation services, needed for the remote control equipment and human comfort shall be independent from the base building HVAC system. The remote control facility shall not be dependent on the base building HVAC system for these services.

**Communications.** Telecommunications services inclusive of voice grade services shall also be independent of the base facility services.

### Auditing and inspection

**Initial certification process.** The reviewing entity shall certify the applicant's compliance with the provisions of this regulation. Certification may include, but not be limited to, physical inspections of the central monitoring facility, review of emergency services detailed above, and inspection of the qualifications of the operating personnel.

**Recurring compliance inspections.** Initial certification notwithstanding, the local reviewing entity shall retain the right of periodic inspections of the remote control facility to certify continuing compliance with all the provisions of the regulations. Certified applicants found in noncompliance shall be given 30 days to rectify the deficiency. Pending a reinspection of the identified deficiency, the remote control facility shall either be recertified as compliant or lose its designation as a certified remote control facility. In accordance with the continuing certification process, the remote control facility shall provide the reviewing entity with a list of all facilities where the applicant performs remote control services.

### Controlled facility

The following provisions relate to the minimum requirements of the facility being controlled by a certified remote control facility. Where variable sensed values are denoted, analog sensors shall be provided which measure the denoted temperature, pressure, amperage, etc. Where status points are indicated, positive indication shall be provided.

#### Methods

Minimum monitoring points

#### Refrigeration equipment

Chillers (200-ton minimum). Start/stop, status, evaporator pressure, condenser pressure, compressor amperage (one phase), leaving condenser water temperature, chilled water supply temperature, chilled water return temperature.

Chillers (200-ton maximum). Start/stop, status, chilled water supply and return temperature, leaving condenser water temperature.

#### Boilers

Steam (75 BHP minimum). Start/stop, status, flame fail, steam pressure.

Hot water (75 BHP minimum). Start/stop, status, flame fail, hot water supply temperature, hot water return temperature.

75 BHP and smaller. No requirement.

#### Air handling equipment

20,000 cfm and above. Supply fan start/stop, supply fan status, return fan start/stop (if so equipped), return fan status (if so equipped), supply air temperature, return air temperature, mixed air temperature (where equipped with economizers). Supply air static pressure (for variable volume systems).

Below 20,000 cfm. Supply fan start/stop and status, return fan start/stop and status (if so equipped), supply air temperature, return air temperature.

### Pumping systems

Applies to primary hot and chilled water, secondary hot/chilled and condenser water pumping systems, where pumps exceed 25 HP. Start/stop and status of all pumps (except stand-by devices). Delivery water and return water temperatures.

### Data collection devices

Data collection devices shall include nonvolatile random access memory for the storage of system parameters. Memory shall include the database of connected points, fail-safe positioning, and battery backup to protect the stored information. Data collection devices shall be UL listed and installed in accordance with the manufacturer's recommendations. Monitoring and control points may be connected to the data collection devices via: hardwired, power line carrier, or radio frequency communication methods in accordance with the NEC, NFPA, IEEE, FCC, and any other government agency having jurisdiction.

### Data transmission devices

Data transmission devices (modems, transmitters, active hubs, etc.) shall conform to FCC regulations for data transmission devices. All data transmission equipment deployed in support of the remote control services shall be dedicated to the remote control system and shall not be used for voice transmissions or other computer system data communications services.

## General provisions

**System maintenance.** All monitoring and control points utilized in the performance of remote control services shall be maintained in proper operating condition in accordance with the manufacturers recommendations. The requirement extends to the installed monitoring and control points, data collection devices and data transmission devices located within the controlled facility, together with all data transmission systems, data processing, and storage systems at the remote control facility. The applicant shall provide evidence to the reviewing entity of compliance with the provisions of this section.

**Electronic watch tours.** All of the provisions of this regulation having been met and certified by the reviewing entity, the applicant may utilize the remote control and monitoring facility to perform engineering watch tours (required by local Code) of the connected equipment electronically.

# NOTES

Chapter

# 6

# Custodial Services

## Types of Cleaning

As an effective facility manager, it is your responsibility to define the types of cleaning that will be required and expected within the facility. This can only be accomplished by obtaining an understanding of the property and its needs through an investigation of the following criteria:

- Types of areas requiring service
- Parallel (or "in common") requirements of the tenants
- Uncommon needs of the tenants
- Budget guidelines for the cleaning program

With a clear understanding of these concepts, you should have the basis for building a cleaning program that will best serve to protect the facility assets and to keep the tenants happy.

### Cleaning program

The first step in creating a clear understanding between the facility manager and vendor is to pre-prepare a list of *specifications.* In the cleaning industry, this is called the list of "specific ACTIONS." It is from this offering of specific ACTIONS that the cleaning contractor will decide: the number of employees required on a daily basis to clean the facility, the amounts and types of supplies and equipment that will need to be present on-site to accomplish scheduled goals, and ultimately, what costs will be charged to the facility to justify the contract. So, as can be seen, your attention to detail in writing the proper specifications will determine your "effectiveness-to-budget" ratio.

In order to get started, let's begin with the fundamental educational process within the cleaning industry. Every cleaning contractor begins by referring to the "Five Steps of Basic Cleaning." These steps (or categories) are as follows:

Step 1. *Trash.* Functions that must occur within the contract scope that pertain to the collection, removal, recycling, compacting, breakdown, and hauling of rubbish.

Step 2. *Ash.* Actions that will occur within the parameters of the cleaning agreement that address the collection, cleaning, and responsible removal of all refuse produced by smoking (whether within the general office environment or within the confines of smoking areas).

Step 3. *Dust.* All dusting tasks that must occur within assigned work spaces to minimize the collection of dust particles on horizontal and vertical surfaces.

Step 4. *Spot cleaning.* Refers to damp cleaning glass and counter surfaces (both horizontal and vertical) to remove the presence of stains, spills and hand prints. Note that this does not fall within the category of dusting because the nature of the action involved requires specific technique and solutions to accomplish this task.

Step 5. *Floor work.* This category encompasses all tasks required to maintain any and all floor types within the serviced property. For example, carpeted floor areas and the different pile types, rugs, vinyl composition floors, stone floors (marble, granite, terrazzo, etc.), ceramic tile, wooden floors, and raised floorings would all need specific maintenance requirements.

It is from this reference guide that facility "specific ACTIONS" should be constructed.

Now it is time to address the concept of "frequency." Each task on your list will require a "times per year" total to determine how often the action will need to be performed. It is important to identify which tasks will require more frequent occurrence from the tasks that will require less attention. By determining these frequencies, you will be in a position to expend the budgeted funs with greatest efficiency. Keep in mind that some of the tasks that will require the lowest occurrence can be the most costly to perform "each time" due to labor burden.

### Frequency definitions

*Daily.* Specifies that this task will be performed every day that the facility will be open for business. Most facilities provide "full service cleaning" for its tenants five days per week either Monday through Friday or Sunday through Thursday. (Of course, porter and matron services are typically on a Monday through Friday schedule regardless.) The term daily indicates the number of days during the year in which a facility will be open for business—typically, 260 days per year for five days per week service or 365 days per year for seven days per week service.

*Weekly.* Indicates that a task is to occur one time per week throughout the areas in which it is specified.

*Monthly.* Tasks that are specified to be performed "monthly" will generally occur one time every 30 days and may be listed as having a frequency of 12 times per year.

*Quarterly.* A "quarterly" frequency task is scheduled to be accomplished four times per year or one time every three months. Certain specification programs may describe this task and then simply list the number 4 next to the description, indicating four times per year as the occurrence schedule.

To outline the remaining, most common, frequency descriptions, the following frequency descriptions are listed:

Three times per week = 156 times per year

Four times per week = 208 times per year

Bi-weekly = 26 times per year

Bi-monthly = 6 times per year

Semiannual = 2 times per year

Annual = 1 time per year

It is important to identify "frequency" of service as an important part of your thought process when building the "specific ACTIONS" for your facility, as this will be the guideline in which the contractor will adhere to for the duration of the agreement. The tasks that are typically underestimated in frequency mainly deal with the up-keep of floor surfaces. Whether it be carpet extraction, hard surface floor maintenance (i.e., stripping and refinishing, spray buffing, stone polishing), or restroom ceramic and grout care, it is important to consider the traffic that will cause the deterioration of these areas and to address them through the specifications accordingly.

The next step in determining the personnel required to maintain your facility is to touch on the subject of "productivity rates." These rates are directly derived from the combination of "specification" and "frequency." Each task that will be accomplished within your facility will have (by industry standards, and job knowledge and experience) a productivity rate individually assigned to it based upon the average derived from having repeatedly performed each task and assessing a productivity value to each duty. This productivity rate is generally based upon "the number of square feet that can be covered by an individual task in one hour." For example, the industry productivity rate for collecting trash in a "typical" office environment is 20,000 $ft^2/hr$ according to the Daniels Associates–engineered performance standards for performing this "general cleaning" function. And just as the task "trash collection" has an industry productivity standard assigned as a general rule, so does each task that will be performed within your facility. Some floor finishing tasks will be accomplished at a rate as low as 200 to 300 $ft^2/hr$ based upon industry standards.

Once each task has been assigned a "productivity value" based upon a "square feet coverage rate per hour," an overall productivity rate can be determined that best suits the specific needs of your facility.

### Staffing

And it is from this "overall production rate" for completing all daily, weekly, monthly, semiannual, and annual duties that the cleaning contractor will most efficiently staff your facility. Without exposing any preconceived thoughts about productivity, as each facility requires different and specific production needs, the following hypothetical observation is presented.

Let's say that ABC Corporation is the single occupant of a 300,000 $ft^2$ (cleanable space) building. The building area usable space is divided into the following categories: Executive Office Space, General Office Space, Hallway and Corridor Space, Restroom Space, Deli Space, Dining Room Area, Kitchen Area, Delivery Dock(s), Fitness Area, Elevators, Main Lobby Areas, etc. Now, once these areas have been identified, a set of specifications will need to be developed to address the specific cleaning needs for each space. It will not be unusual to find that many of the cleaning processes found within the Five Basic Steps of Cleaning (trash, ash, dust, spot cleaning, and floor work) will apply to all of the areas at first writing. However, it is when the special requirements for each area are developed that we discover vast differences in cleaning applications. It is the responsibility of the facility specifications developer(s) to carefully apply the correct cleaning tasks to each area in order that a proportionate amount of required time will be allocated to each space. Once this has been accomplished, each separate area will be once again combined to equal the facility "cleanable" square footage and a production rate will now be calculated for the facility. The following example will substantiate this process:

| | |
|---|---|
| Executive office space | 50,000 $ft^2$ |
| General office space | 210,455 $ft^2$ |
| Hallway and corridor space | 21,000 $ft^2$ |
| Restroom space | 5,000 $ft^2$ |
| Deli space | 500 $ft^2$ |
| Dining room area | 2,800 $ft^2$ |
| Kitchen area | 1,200 $ft^2$ |
| Delivery dock(s) | 1,000 $ft^2$ |
| Fitness area | 2,300 $ft^2$ |
| Elevator(s) | 245 $ft^2$ |
| Main lobby area(s) | 5,500 $ft^2$ |
| Combined cleanable area total | 300,000 $ft^2$ |

As can be seen, each area has been assigned an individual square footage for the amounts of space utilized. As discussed, each area has its own "individuality" with regard to the cleaning process, so the proper set of specific ACTIONS for each area has to be determined as well as the "frequency" these specifications will need to be performed throughout the contract period. Each area has its own "productivity rate" based upon a combination of specifications and frequencies. And, when each area is folded into the total area, an "overall productivity rate" is then determined for the entire facility.

It is from this rate that the general cleaning contractor will propose staffing for the project. The above information has been proven as a worthy insight and illustration for determining productivity levels within your facility. The next logical course of action is to further review the staffing requirements that pertain to identifying job descriptions for the custodial staff that will be responsible for cleaning your building on a daily basis. To illustrate this concept, we once again refer to our hypothetical 300,000 ft$^2$ building, occupied by the ABC Corporation.

ABC Corporation is a company that produces widgets and sells them on the world market. Like many companies, ABC is housed in an office structure with assigned spaces hosting a list of amenities that produce a comfortable working environment for employees and visiting guests. As previously mentioned, ABC Corporation has assigned facility spaces for each area listed above. The facility manager responsible for contracting cleaning services for the building has provided specifications for each area, along with a desired frequency schedule, and has allowed the prospective custodial provider to submit a proposal that outlines staffing for the project. From this list of specific ACTIONS, the contractor has applied industry accepted productivity standards and has established a cleaning rate of 3750 ft$^2$/hr. Keep in mind that this proposal has been tailored specifically to the requirements specified by the facility manager and directly answers the criteria presented. (In other words, production rates will vary from facility to facility). Accordingly, the following formula is described for use:

ABC Corporation building cleanable square footage: 300,000 ft$^2$

Determined overall productivity rate: 3750 ft$^2$/hr

Formula: 300,000/3750 = 80 custodial personnel hours required per day

The general cleaning contractor proposes 80 custodial personnel hours as being the minimum requirement for performing the facility managers work specifications. The next phase of this process is to determine how many hours a cleaning shift will be available (or required) by the facility manager. This process can be performed in two ways for illustration.

### Illustration

Full-time shift: 80 hr/day/8 hr (or full-time) shift = 10 available positions

Part-time shift: 80 hr/day/4 hr (or part-time) shift = 20 available positions

As can be seen, both approaches utilize the same number of hours to clean the facility, but based upon the length of the shift, the number of available positions (to accomplish the exact amount of work) is different. This can become confusing if you are accepting competitive bids and one contractor proposes a full-time shift and one opts for a part-time shift. This can easily be clarified by simply comparing cleaning hours. Do not be confused by the number of personnel, because as demonstrated above, the contractor proposing 20 persons is providing the same amount of cleaning hours as the contractor proposing 10 people.

### Position descriptions

The part of the process has now been reached in which job descriptions need to be assigned to the proposed custodial staff. At this point in time, we have a facility that needs to have general cleaning services, and a staff of 20 persons to perform the custodial duties. The next step is to create position descriptions for each of the cleaning staff and to direct them to individual work areas to execute these duties. To simplify this process, once again refer to the *five basic steps of cleaning,* for it is from the list of specific ACTIONS provided by the facility manager that job assignments will be created. Before moving into specific job categories by position, the following job position titles are listed:

General Cleaner (a.k.a. General Service Worker (GSW), Maid or Matron)

Restroom Cleaner

Floorwork Personnel (a.k.a. Floor Technicians)

Utility Cleaners (a.k.a. Utility Support Cleaners, Zone Cleaners)

Stairway/Stairwell Cleaner

Elevator/Escalator Cleaner

Periodic Cleaners

Window/Glass Technicians

Trash Removal Persons (a.k.a. Trash/Recycle Technicians)

Day Porter (a.k.a. Day Person, Restroom Restocker)

Day Matron (a.k.a. Day Person, Restroom Restocker)

As can be seen, the facility can have a number of different job positions which will be allocated to the general cleaning staff. These staff members will perform specific functions based upon assignment to ensure that the custodial contractor is complying with the job functions and specifications prescribed by the facility manager. To further illustrate this concept, refer to the above hypothetical building occupied by the ABC Corporation.

To recap: The facility manager of the ABC Corporation has accepted a number of competitive bids from the most reputable custodial service providers in the vicinity of the ABC Corporation property. The initial qualifying criteria required for a custodial service organization to be selected as a prospective bidder included proof of financial stability, years in the custodial service industry, references of "like accounts" in square footage and buildout, the ability to provide insurance coverage as required by the ABC Corporation, and demonstration of competent management personnel. Once all bids were submitted, a comparison of "hours to be worked" and "shift length" was performed as well as a comparison of other financial data that was pertinent to each group of prices. It is from this comparison that the ABC Corporation selected the commercial cleaning company offering its services at the most competitive rate. ABC Corporation did not have an obligation to select the lowest bidder, or the highest bidder; but the bidding contractor that offered experience, competitive price, managerial know-how, insurance coverage, and financial strength was chosen. The chosen cleaning organization proposed a part-time shift utilizing 20 personnel at four hours each or 80 hours total dedicated to performing custodial services per night. (Refer to preceding productivity illustration.) It is from this identification of numbers of cleaning staff personnel required for the job that position descriptions can be assigned.

### Position One: General Service Worker

This person is responsible for performing all daily housekeeping and security tasks (i.e., locking tenant doors, turning off lights upon job completion) as stipulated within the specifications and directed by the facility manager. Duties include: trash collection and re-lining of receptacles with plastic liners as needed, general dusting to include daily and periodic high (objects over 60″) and low (objects under 24″) dusting, spot cleaning of horizontal and vertical glass and laminate surfaces with a clean damp cloth, and general vacuuming of carpets and rugs.

In other words, the General Service Worker (GSW) position is responsible for most of the duties found within the "Trash, Ash, Dust, and Spot Cleaning" categories. Therefore, when the position breakouts are demonstrated, the GSW position will consume the majority of the 20 staff members proposed because, as shown previously, the ABC Corporation consists mainly of executive office space, general office space, corridors, and hallways (281,455 $ft^2$).

### Position Two: Restroom Cleaner

This person handles all daily restroom cleaning tasks as required by observation and contract specifications and directed by the facility manager. Duties include: restroom trash collection and re-lining of receptacles with plastic liners daily, damp disinfecting all fixtures and sinks, restocking supplies (towels, bathroom tissue, feminine products dispensers, soap dispensers, etc.), damp disinfecting of all horizontal and vertical surfaces to include partitions and mirrors, sweeping, and mopping (with a germicidal solution) floor surfaces.

### Position Three: Floorwork Personnel (Floorwork Technicians)

The floorwork position is normally assigned to an individual experienced in maintaining various floor coverings and surfaces throughout the facility together with the skills necessary to operate the equipment required to perform this work. A knowledgable floorwork person has insight and background in shampooing, extracting, and spot cleaning carpeted floors as well as being able to perform hard surface floor care tasks such as polishing stone floors, stripping and waxing vinyl composition floors, spray buffing (the technique utilized to spray a fine layer of floor finish onto a waxed floor and buff in order to maintain the luster of that floor on a daily or weekly basis without having to strip the surface), dust mopping (or sweeping), and damp mopping. These duties are generally performed under direct guidance by facility supervisors in accordance with contract specifications and scheduled frequencies.

### Position Four: Utility Cleaners (Utility Support Cleaners and Zone Cleaners)

Utility cleaners are assigned their duties by the facility manager in accordance with the specifications' areas that would not fall under the responsibility of the General Service Worker (GSW) position. Typically, it is the job of the utility person to maintain the normal, day-to-day, cleanliness of areas such as facility lobbies, elevators, escalators, retail areas, fire escape stairwells, building entryways, walkways, and plazas, loading docks, freight elevators, skywalks, basement-level service hallways, and tunnels. This position can also be designated as "Utility Support Cleaner" because the individuals occupying these positions may be called upon to assist GSWs in accomplishing periodic task work that is not performed on a daily schedule, but on a periodic frequency such as high and low dusting. When the utility cleaner is assigned as a support position to accomplish periodic tasks, he or she may choose to accomplish these duties by dividing their assigned work areas into "zones" (thus the title "Zone Cleaner"), accomplishing the periodic tasks schedule in these smaller portions of the assigned work area on a daily basis versus, say, all of a task in the entire work area in one day. To clarify, let's say that our utility support cleaner has been assigned to perform the "weekly," "low-dusting" task and the "monthly," "high-dusting" tasks in a designated work space. This person may choose to divide this work area into four equal quarters (or "quadrants") and execute this work utilizing the following sample schedule:

Monday (perform weekly low dusting in Quadrant 1)
Tuesday (perform weekly low dusting in Quadrant 2)
Wednesday (perform weekly low dusting in Quadrant 3)
Thursday (perform weekly low dusting in Quadrant 4)

At this point, all low dusting is 100% complete in the entire work area.

Friday (perform monthly high dusting in Quadrant 1)

This monthly high dusting schedule is to be completed in one quadrant every Friday until all quadrants have been high dusted for the month. At the beginning of each month this cycle repeats itself.

#### Position Five: Stairway/Stairwell Cleaner

The person or persons assigned to maintaining cleanliness of the facility stairways (tenant internal suite stairs, carpeted public or common area stairs) and building stairwells (fire escape stairs, generally concrete or vinyl tile; used very little or in emergencies) will do so in accordance with the work specifications assigned to these areas. It is common for the Stairway/Stairwell position to also fall under the Utility Cleaner category. This is because many facilities that have stairs do not provide enough stairway space to require a cleaner or cleaners to spend their entire shift cleaning stairwells.

You will generally find the Stairway/Stairwell position to be assigned to a person or persons in very tall buildings, (say, 20 stories or greater), or in a multiple building complex. Other than these two scenarios, stairwell cleaning should fall under the responsibility of the Utility Cleaner position which will also be subject to accomplishing other assignments.

### Position Six: Elevator/Escalator Cleaner

Much like the Stairway/Stairwell Cleaner, the Elevator/Escalator Cleaner will typically fall into the Utility Cleaner position description as well. The facility would have to have an extremely large number of elevators and escalators to occupy the entire shift length of one cleaner. It is possible for this to happen, but more often than not elevators and escalators will be cleaned by a Utility Cleaner. Once again, in the situation where a facility has an extremely large number of elevator cabs, it is possible to create a single position, or multiple positions, dedicated solely to this type of cleaning. The main responsibility of the Elevator/Escalator Cleaner is to clean the elevator cabs in their entirety to include all walls, ceilings and fixtures, metal doors, metal railings, door tracks, button push plates, escalator tracks, rubber hand rails, glass, decorative metal, etc.

### Position Seven: Periodic Cleaner

In custodial work terms, “periodic work” or “periodics” are jobs that are outlined within the facility specifications as required to be accomplished on a frequency schedule less than daily. These duties are generally scheduled on a weekly, monthly, quarterly, biannual, or annual basis. Therefore, many facility custodial managers will create full positions dedicated to accomplishing these tasks on a periodic schedule which complies directly with building specifications. The role of the Periodic Cleaner may also fall under the Utility Cleaner category depending on facility size. The Periodic Cleaner position may have the responsibility of performing duties such as light lens cleaning, which typically happens on a quarterly to annual frequency.

### Position Eight: Window Glass Technician

Many facilities are built with a skin consisting of windows that are set in sills. This window glass has two sides: the exterior side and the interior side. It is usually the interior side of this glass that is directly addressed by the custodial contractor in accordance with facility specifications. Along with this interior/exterior glass, many buildings have been furnished with many decorative accessories that have either glass or plexiglass panels such as partitions, glassed in conference areas, glassed in offices, and computer room areas. These glass panels require daily cleaning to remove spots and fingerprints, as well as periodic cleaning of the panels in their entirety to remove dust that will accumulate on the upper portions of these panels that are not subject to fingerprints and smudges. It is therefore the responsibility of the custodial contractor to address the glass cleaning issue by either assigning a sole position, or positions, to clean and maintain window glass, or to schedule this task to be executed by the Utility Crew. It is highly unlikely that any part of the cleaning crew will have responsibility for cleaning the exterior window glass higher than the first story. It is generally common practice for a separate, specialized window crew or service to be responsible for maintaining all exterior window glass higher than that of the first story of the structure. This also will remain true for interior skylights, atrium glass, and lobby glass that is higher than the first floor elevation where either scaffoldings, platforms, or lifts are required. In this case, special insurance requirements will also need to be addressed due to liability issues.

### Position Nine: Trash Removal Cleaners (Trash/Recycle Technicians)

Trash Removal Cleaners (not to be confused with those individuals assigned to trash collection) are the personnel responsible for taking collected and bagged trash to its final destination at the facility for pickup and transporting (or hauling) to the dump or recycling vendor. These persons can also be classified as "Trash/Recycling Technicians" if the job requires that they become proficient in the sorting of materials that have the potential for recycling such as glass, plastic, paper, and cardboard. Many facilities have a recycling program in place where the building inhabitants have received instructions to place reusable materials in bins that have been labeled for the collection of those materials only. The responsibility of trash collection will generally fall under that of the General Service Worker (GSW).

The GSW will typically gather all trash in 44 gallon containers or "brute barrels," which are large plastic collection buckets on wheels for easy transport. These containers are generally lined with a large plastic bag that is tied off when full. The GSW will then place the full bags at a gathering point somewhere in their work area (near an exit door or in the elevator lobby). Then, at some point during the shift, a Trash Removal Cleaner will make rounds with a very large collection receptacle, say a 1.5 cubic yard buggy, and gather the bagged trash. This trash is then moved to the area on-site designated as the final collection point before transport to the dump. This area is typically a loading dock, or platform, which provides a dumpster or trash compactor. In some instances, the facility will provide a number of different variations for trash to be sorted. This might include a number of different dumpsters labeled as "Paper Only," "Plastic Only," "Glass Only," or "Cardboard Only." It is therefore the responsibility of the "Trash/Recycle Technician" to properly sort these items in order to provide the greatest possibility for the most efficient recycling of these items.

It is at this point in our position descriptions where a distinguishable line can be drawn that separates the facility cleaning staff into two entities; the "night staff" and the "day staff." Both of these facets play an integral role to comprise the whole with regard to providing full service within a facility. Where the inhabitants of the facility rarely come in contact with the evening general cleaning staff, the opposite is true for the day staff. The day porters and day matrons act as the "glue" to keep the entire cleaning operation together and generally grow to become relied upon by building tenants to provide continuity to the cleaning service throughout the day. It is the important responsibility of the day personnel employed to maintain the flow of service during normal building hours. Without the day staff (and assuming that the facility is of the size where day services are required), it is possible that the efforts of the night staff could fall short in a number of areas such as restroom supply restocking due to daytime usage, cleaning of spills, and in the instance where a large quantity of trash has been produced and is in need of removal. It is from this information that the following positions have come to play an integral role to the success of many facility custodial service providers.

### Positions Ten and Eleven: Day Porter/Day Matron (Day Person, Restroom Restocker)

The Day Person position has become vital to the continuous flow of custodial service within many facilities. It is not uncommon for many tenants to rely upon the daytime cleaning crew for a number of crucial services that make use of the work place a pleasant experience. For example, the Day Crew will generally report to their assigned positions at least one hour prior to normal facility work hours, and sometimes earlier. Day personnel will ensure that common areas and entryways are clean and free of debris. These day positions are also responsible for keeping plazas and walkways safe by removing any standing water, snow, or spills that might be present prior to normal work hours and that might occur throughout the day. Once a normal rotation has taken place to make common areas and entryways clean and safe, a scheduled check of facility restrooms takes place. This is one reason why day persons might also be titled "Restroom Restockers." During the morning hours, normal building activity will deplete restroom supplies such as hand towels, bathroom tissue, and hand soap. Therefore, it is the standard schedule of the day porter/day matron to check and restock these supplies as needed to avoid any of these items from being totally stocked out. The only real distinguishable difference between the job of the Day Porter and Day Matron is that it proves less of an inconvenience to facility tenants when no interruption in restroom services occurs due to opposite gender restocking. Because of this, "Day Person" has grown to become the industry-accepted title for this position. It is not uncommon to see facility day custodial personnel assisting the facility manager in performing a number of tasks when not attending to normally scheduled custodial duties.

It is at this point in the discussion that the above presented information will be summarized. It is assumed that the information regarding "general cleaning," "specifications or specific ACTIONS," "the five steps of basic cleaning," "frequency," "productivity," "cleaning rates," "cleaning hour comparison," and "job descriptions" have been presented in a format that is easy to comprehend and use. The end result of the above presentations provides a means for creating job descriptions for the ABC Corporation staff that will allow the facility to be cleaned utilizing the most efficient methods available today.

The ABC Corporation staff consists of 20 general cleaning personnel. Each of these people will be given a job assignment in accordance with building specifications. From the information above it is known that their job titles could be any one of the listed categories "general service worker," "restroom cleaner," "floorwork technician," "utility cleaner," "stairway/stairwell cleaner," "elevator/escalator cleaner," "periodic cleaner," "window/glass technician," or "trash removal person." Each custodian's eligibility for any of these positions will be based on previous job experience. The following diagram shows the building staff by position:

| | |
|---|---|
| Executive office space | 50,000 $ft^2$ |
| General office space | 210,455 $ft^2$ |
| Hallway and corridor space | 21,000 $ft^2$ |

| | |
|---|---|
| Deli space | 500 $ft^2$ |
| Dining room area | 2800 $ft^2$ |
| Kitchen area | 1200 $ft^2$ |
| Fitness area | 2300 $ft^2$ |
| *Combined square footage* | *288,255 $ft^2$* |
| Required "General Service Workers" | 15 |

| | |
|---|---|
| Restroom space | 5000 $ft^2$ |
| *Total square footage restrooms* | *5000 $ft^2$* |
| Required "Restroom Cleaning Staff" | 1 |

| | |
|---|---|
| Main lobby area | 5500 $ft^2$ |
| Elevators | 245 $ft^2$ |
| Delivery dock | 1000 $ft^2$ |
| *Total "common / entryway" area* | *6745 $ft^2$* |
| Required "Elevator, Stairway, Lobby, Dock" Cleaning Staff | 1 |

The remaining staff will be responsible for the remainder of the work which is based on task and not square footage.

| | |
|---|---|
| Required Trash Removal Staff | 1 |
| Required Utility Cleaner, Glass Technician, Periodic Cleaner | 1 |
| Required Floorwork Technician | 1 |
| Total cleaning staff | 20 |

As demonstrated in the above staffing determination example, some of the job positions require proficiency in a number of skills to provide the most efficient staff allocation. The most important factor in the proposed staffing profile is the overall effectiveness of these individuals as they are positioned. This will generally be revealed within the first few weeks of their presence within the facility, so it will be critical to gauge the satisfaction of the tenants during this time. The feedback of the building inhabitants will be a crucial link in allowing the custodial contractor to make the proper adjustments (if needed) with regard to efficient job allocation.

### Quality control (QC)

*Quality control* is an aggregate of functions designed to ensure adequate quality in manufactured products by initial critical study of engineering design, materials, processes, equipment, and workmanship followed by periodic inspection and analysis of the inspection to determine causes for defects and by removal of such causes.

As it pertains to the General Cleaning Industry, quality control is a term that has parameters that encompass a broad scope of smaller entities that make up the QC program as a whole. The facets that "should" comprise a comprehensive QC program in the cleaning service industry are:

Adherence to specifications + cleanliness + communication + execution + personnel + response + training + follow up = "quality"

Experience indicates that any incident that can occur within your facility will fall into at least one of the above categories. The factor that will determine the success (or failure) of the custodial service being provided at your facility will rely solely on the execution of *all of the above.* With this in mind, the *Quality Rating Program,* which has been implemented at the ABC Corporation facility, will be described.

The ABC Corporation is the sole tenant of a building that is approximately 15 years old. The building has been well maintained mechanically and has been renovated and remodeled recently. It has been the goal of ABC to provide a pleasant working environment for its inhabitants and their guests, and an excellent job has been done to keep the interior and exterior of the building looking neat and clean. The recent renovation allowed ABC to update carpets, wall coverings, some furniture, and fixtures in order that the building can maintain its Class A status. So, as can be seen, it is the goal of ABC management to provide the vehicle for custodial quality by instilling a sense of cleanliness and pride within the facility boundaries. The facility manager made quality control a major issue when choosing their custodial service provider. ABC expects quality from each of its vendors.

During the final selection process, ABC was impressed by the quality control program that was outlined by the successful custodial service bidder. It is important to keep in mind that the custodial service that was awarded in the cleaning contract is able to provide quality control in its service effort while still maintaining competitive pricing. This program is now in effect at the facility site and is as described following.

**Step one: communication.** It was the objective of the successful custodial service company to establish lines of communication with the ABC Corporation management office and the facility tenants from the beginning of the relationship. A number of brief meetings were attended prior to commencement of services to discuss issues ranging from: security, keys, special working hours, special facility needs, areas where tenants may work late, heavy traffic areas, possible specific tenant needs, day porter duties (scheduling), and project management and supervisory requirements. A communications log will be present at the site to track cleaning requests and comments in order that total customer satisfaction is achieved.

**Step two: training.** Once a crew was established to clean the ABC site, a staff training plan was drafted to demonstrate hazard communication (Hazcom) training; proper use of HEPA filter vacuums; special cleaning requirements within their assigned working areas; an overview of facility specifications which will be performed within individual workstations; the proper use of tools and equipment; the proper use of cleaning chemicals; and correct maintenance practices of all supplies, equipment, and storage areas.

**Step three: start-up.** A team was assembled by the successful custodial vendor to be on-site at the ABC Corporation facility during the first few days of start-up. This team consisted of management personnel who were assigned to a number of individuals on the cleaning crew for the purpose of conducting extensive training to those persons. The members of the phase-in team are responsible for checking a specific assigned area for these days in order that all specifications be addressed. Should any deficiencies be detected, further training will be provided.

**Step four: equipment.** An extensive equipment list was provided to demonstrate the types of innovations that will be introduced on the ABC Corporation site. Vacuum cleaners equipped with HEPA (high efficiency particulate air) filters will be utilized so that dust particles that are removed from carpets will not be recycled into the air through vacuum exhaust. All vacuum bags will be contained within compartments featured on this equipment. Equipment will be of sturdy construction by manufacturers with an outstanding reputation in their field. The custodial contractor will have open lines of communication with equipment representatives who will answer any questions that may arise with regard to proper maintenance and operation.

**Step five: monitoring.** The Quality Control program will be monitored through a number of on-site visits each year to gauge the success of the cleaning operation. Adherence to a number of criteria will be judged in order that an overall picture of quality can be produced. The graded categories include: Overall Account Cleanliness, Management Satisfaction, Equipment Maintenance, Adequate Supply Stock, Tenant Satisfaction, Log Entries, Adherence to Specifications, Periodic Scheduling, Accurate Billing Records, Cleaning Personnel Satisfaction, On Time Payroll Deliveries, Response to Service Requests, Competence of Site Cleaning Management and Supervision, and Compliance regarding Health, Safety, and Hazard Regulations.

**Step six: reporting.** The final step in providing an all encompassing quality control program is the proper reporting of inspection results. Therefore, the custodial company has implemented an information system where a "Report Card" will be distributed to Facility Management Personnel, and Cleaning Management and Supervision so that deficiencies may be corrected. As a result of these reports, rewards will be issued to individuals demonstrating outstanding efforts and retraining will be conducted as needed. As can be seen, a number of factors comprise an effective quality control effort. Communication is the key in making the entire system work. In all cases, the failure to express cleanliness objectives by the facility manager to the custodial contractor will expose weaknesses in the program. By the same token, failure for the custodial company to respond correctly to customer requests, and within an adequate amount of time, will also undermine established quality objectives. Communication is the fuel that drives the quality machine.

### Waste management program

Requirements for different types of facilities with regard to establishing a successful waste management program within a facility may vary, so we go to the ABC Corporation building once again to present the final model. A number of scenarios, terminology, and equipment choices will be included which can be utilized to effectively address the management of waste collection and removal from the ABC Corporation.

The procedure for collection of rubbish on any site is, in essence, very much the same. Generally, the entire idea is to collect, remove, and transport the waste produced on-site to the dump. It is when the specific goals of the facility are addressed and Environmental Protection Agency (EPA) compliance issues are factored in that very distinguishable differences might be brought to light. These differences can be based upon a number of different factors including: economics, recycling, and philosophy (with regard to the environment). It is at this point that a knowledgable custodial contractor can assist in producing the best waste removal situation for the project. The first step begins with the contractor listening to the needs of the facility manager, for it is from these initial requirements that the program will be established.

Without listing specific state, local, and EPA requirements, rules, and regulations due to the vast scope pertaining to these issues, it is assumed that the facility manager and the custodial contractor are aware of specific requirements for the dumping of waste. Once again refer to the ABC Corporation.

The ABC Corporation has discussed their facility needs for waste disposal at the property. These needs include providing a comprehensive waste management program that will address all issues regarding the collection and disposal of all trash. It is the goal of the ABC Corporation and its inhabitants to recycle as much waste as possible. Between the facility manager and the custodial contractor an interoffice e-mail has been prepared and sent to each of the building tenants outlining the recycling program. Because the ABC Corporation is a single inhabitant of the building, the e-mail was the most efficient way to inform its employees of the program, while saving any paper that might have been utilized for this process. As you can see, ABC is very serious about reaching its recycling goal. The e-mail memo lists a number of methods that will be adhered to by its workers to ensure that the maximum amount of recyclable materials will be eligible for reuse. Some of the instructions are as follows:

1. Each employee is provided with two office trash cans at their desk. One receptacle will have the recycling symbol clearly posted. The recycling bin is for all paper waste that is produced at each desk area. The second waste can is for any other trash that is not specified as being eligible for the recycling program.
2. Any consumption of food or drink (where drink containers produce waste) is prohibited at employee desks to prevent accidental contamination of recyclable paper. This practice is also used to minimize the amount of plastic trash can liners that will need to be replaced due to wet trash. (Dry liners that are not soiled or damaged can be used over and over.)
3. Large (95 gallon) containers on wheels are located in the freight elevator lobbies on each floor. Each employee is

responsible for carrying and disposing the contents of their recyclable (dry) paper waste containers one time per week into these big containers. When full, these 95 gallon containers are moved to the area designated as recycle goods storage. These items will then be loaded and hauled to the recycle vendor.

4. A number of clearly labeled receptacles are located in the kitchen and cafeteria areas that will direct employees where to dispose of plastic containers, glass containers, and aluminum cans. All other wet trash packaging (i.e., sandwich wrappings, sandwich boxes, Styrofoam meal containers) is disposed of in one barrel which is also located in the kitchen and cafeteria area. This barrel does not contain any items considered for recycling. It is critical that no wet trash come in contact with the paper waste that will be recycled. This contact could contaminate the paper trash, possibly rendering it useless to the recycling process. All collection bins are removed when full and the contents will be hauled for reuse.

As can be seen, the ABC Corporation is determined in its effort to introduce a successful recycling program that encompasses all eligible goods.

The next part of the waste management process is to provide a list of equipment that will be necessary for accommodating the recycling goal of the ABC Corporation. This equipment is essential in providing to the program the means for proper execution. The waste containers must be the proper size to hold the amounts of waste that will be produced at the facility. This equipment must also be of sturdy construction and designed well in order that transport of full containers will be manageable. With these goals in mind, we offer the following list of commonly proposed equipment.

1. *Ninety-five gallon paper collection containers with wheels and lids.* These large containers are constructed of a very durable plastic. The objective is to place these large receptacles in an area where employees can dump paper wastebaskets from their desks. These containers roll off much like a two wheel furniture dolly (tilt and roll) for easy

transportation to and from the areas in which they are located. They are available in a variety of colors for easy identification.

2. *Fifty gallon containers with wheels and lids.* These medium–large containers are constructed of durable plastic that can be easily cleaned by hosing the entire unit. The objective is to place these receptacles in large kitchen and cafeteria areas where employees can choose the correct receptacle (as labeled) for disposal of glass, plastic, and aluminum packaging. These containers are built with design features that include lids and furniture dolly style (tilt and roll) wheels that allow for easy transport to and from the areas in which they are positioned. A number of colors are available to differentiate these receptacles from one another.

3. *Twenty-eight gallon containers with wheels and lids.* These square containers are available in a variety of colors and feature a four-wheel base and lids. Constructed of durable plastic, the wheels swivel in order that these units can be easily moved in any direction. These 28 gallon units as described can be placed in smaller kitchen areas where 50 gallon receptacles would be too large, or where smaller amounts of recyclable trash will be produced.

4. *Five gallon "We Recycle" deskside containers.* Five gallon "We Recycle" deskside containers are recommended to be stationed at each employee work area. These receptacles are constructed of durable plastic and have the recycling symbol (three arrows forming a continuous circle) clearly posted on them; they are molded in recycle-blue plastic for easy identification. It is the intention that only paper produced at the desk site be placed in these cans. Once they are full, employees can easily carry their can to one of the 95 gallon paper collection containers located near the work area.

5. *Five gallon waste receptacle.* These 5 gallon waste receptacles are positioned at the employee deskside, adjacent to the "We Recycle" container. A plastic liner is placed inside this container, as it is the intention of the company for employees to place any nonrecyclable waste in these containers. These containers are constructed of durable plastic

and can be furnished in a variety of colors. Usually cast in beige or black plastic, these waste cans are emptied nightly by the cleaning crew and re-lined as necessary.

6. *Forty-four gallon "brute barrel" containers.* Also referred to as "janitor carts," 44 gallon "brutes" are probably the most recognizable piece of custodial equipment in a building. Each staff member assigned to general cleaning will receive a brute barrel as standard issue equipment for the collection of deskside trash in their assigned work area. These barrels are constructed of durable plastic and are situated on five swivel wheels for easy mobility in all directions. It is standard practice to line these collection carts with large plastic liners similar in size to leaf and lawn bags. The General Service Workers (GSW) will empty all nonrecyclable desk side and kitchen trash into these large, lined containers nightly for disposal. This trash is hauled directly to the dump (or similar site) due to the nonreusable nature of this trash.

7. *1.5 cubic yard "Tilt Truck."* Constructed of durable heavy duty plastic, the 1.5 $yd^3$ tilt truck (or larger, up to 2.5 $yd^3$) is a large buggy usually gray in color. These carts are manufactured with two fixed rear wheels on an axle and two heavy-duty swiveling front wheels for easy maneuverability. Once the GSWs have collected desk side and kitchen nonrecyclable trash, it is general practice for them to place this bagged and tied trash in a gathering area (typically a freight elevator lobby) for removal from their work area. The person or persons on the custodial crew assigned to trash removal load these "tilt trucks" with the nonrecycle trash collected in the 44 gallon brutes and transport to the building area containing trash dumpsters or compactors. It is the intention of the custodial company to be able to move large amounts of rubbish to the trash docks in the large tilt truck containers each trip, creating a most efficient operation. These containers are designed to be tilted at the edge of a dock platform for easy unloading or dumping, thus the word "tilt" in the description. Tilt trucks are not motorized or mechanical and are pushed much like an oversized

shopping cart with fixed rolling wheels in the rear and swiveling wheels in the front.

At this point, the equipment has been discussed that is located within the confines of the ABC Corporation building and the procedure that has been adopted as policy for waste collection. It is at this juncture that the removal of recyclable and nonrecyclable trash from the Trash Dock area of the facility to the final dumping or recycling site will be discussed. This removal procedure is set up on a schedule by the company or companies contracted to haul recyclable and nonrecyclable waste.

The subject of removal of these waste items from the ABC Corporation building is accomplished as shown following. Go to the waste collection area of the facility. In this instance, it is the loading dock area of the building that features direct street access for large trucks to pick up and deliver materials at the site. In a conversation between the facility manager, the custodial contractor, and the waste collection service, it has been agreed that a number of large metal dumpster containers will be provided to the building for the gathering of waste and recyclable goods. Each dumpster bin will be clearly labeled as to what items will be placed in each bin exclusively. To further reinforce this effort, these bins have been color coded for identification of contents as well as posted in a number of languages for the same purpose. The custodial contractor has implemented a training program that details these criteria and will designate individual persons to place separated materials into these containers. No person will be placed in this job category who has not been sufficiently instructed in the handling of the facility waste materials. It has also been determined that the waste collection service will handle the removal of all facility waste including recyclable goods. Therefore, a pickup and removal schedule has been set up as follows:

1. The waste collection contractor will service the account every day, five days per week. Since the amounts of the

various types of trash (i.e., paper, aluminum, plastic, glass, and general waste) are separated by the building tenants and placed in designated cans, the need to haul any one of these categories of waste daily is eliminated. (There just won't be enough volume of any one item produced for this to be required).

2. The by-product that accumulates the largest amount of waste is paper and paper products such as corrugation. This waste needs to stay dry in moving from the ABC Corporation site to the recycle vendor's plant. Therefore, a sizable vehicle that hauls dry paper goods only will be required to service the account one day per week. A number of vehicles can be utilized for this process as long as the loading compartment or trailer provides adequate cover for the paper product within.

3. A second vehicle will be utilized to remove the remaining recyclable products at the site (i.e., aluminum cans, plastic containers, glass containers, and general waste). Since these items typically produce equal amounts of moisture from their contents, it is not necessary to provide separate hauling vehicles (or cover for that matter). However, it is still important to prevent commingling of these products from the ABC Corporation site to the recycling facility. This can easily be accomplished by placing large (clear) plastic liners in the various collection bins where tenants place their recyclable goods. (The clear plastic liners provide an easy identification method in order that the waste collection contractor can quickly determine the contents of each load). Also, when these goods get to the dump-site, they can be separated quickly from one another as the clear liners afford an easy view of their contents. This process can save the trash collection service valuable time and money as individuals are not required to "hand sort" the trash as this has already been accomplished at the building site.

# NOTES

# NOTES

Chapter

# 7

# Landscaping Services

## General Responsibilities of the Landscape Contractor

It should be the responsibility of the landscape contractor to furnish all labor, equipment, and materials necessary to perform all specified landscape maintenance tasks. All equipment should be of such type and condition as needed to effectively perform the task intended and to avoid any unreasonable hazards or dangers to the properties, occupants, and pedestrians. Equipment should be maintained in good condition and should not produce excessive noise or noxious fumes when operated under normal conditions. Some communities have noise restriction ordinances. These restrictions may apply to mowers, leaf blowers, and other landscape equipment. Check with your local jurisdiction to determine if any restrictions apply.

Personnel employed by the contractor should be thoroughly trained using current and accepted horticultural practices. Trained personnel are less likely to make mistakes and have accidents. Personnel should wear company uniforms at all times as supplied by the contractor. Uniforms help identify employees and add a professional appearance.

There should always be a supervisor, or foreman, on the property to direct all contracted personnel and maintenance operations. Supervisors should report to the facility manager and building manager(s) during each visit to verify their presence and provide an update on the status of the grounds. The supervisor should also leave a work ticket with the facility manager that includes information on tasks completed during the visit. If facility management personnel are not on-site, the contractor should leave the work report in a designated location on-site or mail the pertinent information.

The landscape contractor should comply with any applicable codes and building regulations. This information should be communicated to the contractor prior to the first visit to avoid unnecessary problems. For instance, many facilities have restrictions on mowing hours. If mowing should not occur before 7:30 A.M., this is important to communicate to the contractor to avoid tenant complaints.

All existing and new plant material should be replaced at the contractor's expense if the death or damage of the plant was caused by negligence or a direct act by the contractor. Plant material should be replaced by the owner if the death or damage occurred by natural phenomena outside of the contractor's control. All replacement plant material should be in excellent health and acclimated to the area. It should be a form and type indicative of the species and should be of the same size, type, and form as the plant being replaced.

## Selecting a Landscape Contractor and the Bidding Process

When selecting a landscape contractor several contractors should be interviewed before making a final selection. Cost should not be the only determination in the selection process. Landscape contractor abilities and competence vary widely, and it is important to obtain an objective assessment of the experience and skills that a contractor offers.

The contractor chosen should have the appropriate equipment for the job, should be professional in all respects, and should be appropriately trained. Look for a contractor that listens to your ideas and problems. Ask for references, visit the company's offices, and visit properties they are currently maintaining.

A contractor that is not appropriately trained will likely lead to a frustrating situation where both simple and more complicated maintenance tasks fail to be properly performed. Over time, this will degrade the appearance of the property and will invite additional problems due to the contractor's inability to properly assess plant diseases, identify pests, prune properly, and perform other important landscape tasks.

In order to make apples-to-apples comparisons on different proposals from different contractors, develop guidelines that each contractor must follow for bid submission. Since each property may have different landscape needs and may require different levels of maintenance, it may be best to have a landscaping professional examine the property to analyze its maintenance requirements. Based on this analysis, a comprehensive maintenance specification can be developed along with a list of each task necessary to maintain the landscape and the number of occurrences per task (see Fig. 7.1).

Once the maintenance specifications and any other relevant expectations have been identified, a mandatory pre-bid conference should be called to walk the potential contractors through the property. This meeting should emphasize any special requirements of the property and its boundaries. Any specific questions regarding the landscape can be addressed at this time. In this manner each contractor is bidding on the same exact set of specifications, tasks, and number of occurrences. If this is not done prior to bidding, contractors are bidding on what they think needs to be done, not necessarily what the facility manager wants. By spending the time up front, costly extras and misunderstandings can be avoided.

## Lawn Maintenance

Lawn areas require a consistent maintenance program to stay healthy. If lawn areas are neglected for only a couple of

| | NO. OF OCC. | COST/ OCC. | EXTENSION |
|---|---|---|---|
| **A. TURF MAINTENANCE** | | | |
| 1. Mowing | 28 | $______ | $______ |
| 2. Leaf Removal | 8 | $______ | $______ |
| 3. Edging | 28 | $______ | $______ |
| 4. Fertilizing | 3 | $______ | $______ |
| 5. Weed Control | | | |
| a. Pre-Emergent | 2 | $______ | $______ |
| b. Post-Emergent | 2 | $______ | $______ |
| 6. Pest Control | 2 | $______ | $______ |
| 7. Core Aeration | 1 | $______ | $______ |
| 8. Mechanical Slit-seeding | 1 | $______ | $______ |
| 9. Lime (if necessary) | 1 | $______ | $______ |
| 10. Dethatching (if necessary) | 1 | $______ | $______ |
| | | SUBTOTAL | $ ______ |
| **B. PEST MANAGEMENT** | | | |
| 1. Monitor and Control All Pests and Diseases | 39 | $______ | $______ |
| | | SUBTOTAL | $______ |
| **C. MAINTENANCE OF TREES, SHRUBS AND OTHER PLANTINGS** | | | |
| 1. Edging | 6 | $______ | $______ |
| 2. Mulching | 1 | $______ | $______ |
| 3. Weeding | 28 | $______ | $______ |
| 4. Clean-up | 40 | $______ | $______ |
| 5. Pruning | | | |
| Trees | 1 | $______ | $______ |
| Shrubs | 10 | $______ | $______ |
| Ground cover | 10 | $______ | $______ |
| 6. Fertilization | | | |
| Trees | 1 | $______ | $______ |
| Shrubs | 2 | $______ | $______ |
| Ground cover | 2 | $______ | $______ |
| Annuals and Perennials | 4 | $______ | $______ |
| 7. Monitor and Adjust Irrigation | 19 | $______ | $______ |
| 8. Manual Watering | As needed | $______ | $______ |
| | | SUBTOTAL | $______ |
| | | TOTAL | $______ |

**Figure 7.1** Bid form for apples-to-apples comparisons from contractors.

seasons, they will decline. A maintenance program should consist of proper mowing, pest and weed control, fertilization, watering, aerating, and overseeding. This program will encourage the growth of healthy turf, which in turn crowds out unwanted weeds.

There are two basic types of grasses: cool season grasses and warm season grasses. Cool season grasses can withstand cold winters, but most do poorly in hot summers. Cool season grasses are used mainly in northern areas. Warm season grasses are lush during hot weather and go into dormancy when temperatures are below freezing. A reputable contractor, nursery, or local extension service can determine what type of grasses will thrive in your area.

### Mowing and trimming

Mowing should begin in March or April at intervals of five to ten days (maximum) between mowings. Mowing should be done frequently enough so that no more than one-third of the leaf area is removed at one time. This will help the turf to develop a more extensive root system and withstand environmental stresses. The optimum height for turf is determined by the type of grass. Horizontally spreading grasses are typically cut shorter than vertically growing grasses. Frequent mowing tends to produce a finer-textured turf, since cutting frequently stimulates new growth. If the turf is neglected and becomes too tall, the growth becomes coarse and may produce seeds. Mowing the turf too short will cause the grass to dry and burn, which will allow weed seeds to germinate.

Mowing should be done in alternate directions at least every four mowings. This will eliminate ruts and a striped or streaked look. Mowers should be well maintained and cutting blades kept sharpened at all times to prevent tearing of the leaf blade.

Litter and debris should be removed from all lawn areas prior to mowing. Clippings can be left in the turf area. This is more cost-effective, since the clippings put nutrients back into the soil. If, owing to long periods of wet weather, the turf becomes excessively long, the clippings should be removed, since they can form a mat on the turf and shade out and kill the grass. In areas where there is concern that clippings are unsightly, such as entrances, clippings can be bagged, or raked, and removed. Mowing should be done in such a way that clippings are not blown into the shrub beds and tree rings, since this can be unsightly.

Areas around posts, signs, buildings, and trees should be trimmed at the same height as the lawn. Lawnmowers and string trimmers should not be used at the base of trees and shrubs, since they can cause damage to the base of the plants. The contractor should be responsible for replacement and/or repair of plants, irrigation equipment, exterior lights, signs, posts, fences, automobiles, paved surfaces, building exterior, and any other ground structures damaged by their activities while working on the property.

All sidewalks, pathways, and curbs should be blade-edged on a regular basis. Shrub beds and tree rings should be edged to maintain a 2″ vertical edge between turf and mulch areas. Grass and weeds should be continuously removed from cracks and expansion joints in all walks and curbs. Clippings and all natural debris should be cleaned off all paved areas after mowing and edging.

#### Weed control and pests

Weeds are simply plants that grow in the wrong place. There is no such thing as a weed-free lawn, but with proper control weeds can be minimized. To minimize weeds, the following recommendations are generally applicable:

In early spring, when daytime air temperatures reach 55° to 60°, a broad-spectrum, pre-emergent (applied before weeds emerge) herbicide that controls both noxious grasses and broadleaf weeds should be applied to all lawn areas in accordance with the manufacturer's recommendations. Additional applications of pre-emergent weed control may be necessary to effectively control all weeds.

In late spring and again in early fall, when daytime air temperatures are not above 80°, the contractor should apply a broad-spectrum, postemergent (applied after weeds emerge) herbicide to control all weeds. The presence of certain weed species that are difficult to control may require additional applications of herbicides.

The contractor should regularly monitor all turf areas for insect, disease, and weed infestations; and treat as needed. The contractor should be responsible to replace all turf areas damaged as a result of pest and disease problems with sod. The sod should match the surrounding healthy turf.

### Fertilization

Soil fertility is one of the major considerations in any program of lawn management. A healthy lawn requires a soil that is fertile from year to year. Since grass can quickly deplete soil of essential nutrients, the nutrients should be added into the soil on a regular basis. The essential nutrients for turf areas are nitrogen, phosphorus, and potassium. Nitrogen is critical since it stimulates leaf growth and keeps turf green. Phosphorus is needed for the production of flowers, fruits, and seeds; and to induce strong root growth. Potassium is valuable in promoting general vigor and increases the resistance to certain diseases. Potassium also plays an important role in sturdy root formation.

Fertilizer designations contain three number such as 10-6-4. The first represents the percentage of nitrogen, the second the percentage of phosphorus, and the third the percentage of potassium. All fertilizers list the essential nutrients in this order.

While these nutrients are critical to a healthy lawn, excessive use and application of fertilizers is unnecessary and damaging to both the lawn and the environment. For this reason, apply only the minimum amount necessary to achieve a healthy lawn area.

Soil pH is critical to growing a healthy stand of turf (or any plants). Soil pH is the acid-alkaline balance of the soil. The pH scale divides the range of alkaline and acid materials into 7.0 points. The middle value of 7.0 is neutral, marking a balance between acidic and alkaline soil values. Some plants thrive best in neutral conditions, while others prefer a more acid or alkaline soil. Lawn grows best at a pH of 6.0

to 7.0; consequently, to have a healthy stand of turf it is important to make sure the pH is correct. The pH can be changed by adding lime if the soil is too acid or by adding sulfur if the soil is alkaline.

A soil test will give the pH as well as the level of nutrients available in the soil. From the soil test results a contractor can determine the amount of fertilizer to be applied and if an application of lime or sulfur is necessary.

### Dethatching and aeration

Many soils can become compacted over time. When this occurs, nutrients and water have difficulty in penetrating to the root zone. This causes the root zones to become shallow, which in turn causes the plants to dry out quickly. Aeration can help reduce compaction. Aeration is a method whereby holes are punched into the turf to allow moisture, oxygen, and nutrients to penetrate the soil and reach the root zone. Aeration holes should penetrate 2″ to 4″ into the soil.

Thatch is a layer of dead grass and other organic matter that accumulates between the surface of the soil and the grass blades. When the thatch depth reaches $1/2$″ the thatch should be removed. Cool season grasses should be dethatched in the fall. Warm season grasses should be dethatched in the spring.

### Overseeding

Once a year turf should be overseeded to reestablish the lawn in bare or thin areas. Overseeding is most successful in the fall. The seed should be planted early enough so that the grass has enough time to get established before the cold weather sets in. The next best time to overseed is in the spring after the first frost and before it gets too hot.

## Maintenance of Trees, Shrubs, and Other Plantings

### Mulch and weed control

Mulch regulates soil temperatures, insulates plant roots from temperature extremes, reduces water loss from the soil surface, and minimizes the time and labor required to maintain the garden by minimizing the germination of weed seeds. The most important function of mulch is moisture retention. Mulch allows water to percolate through and protects the soil from the drying effects of the sun.

Mulch is available in a number of forms both organic and inorganic. Organic mulch is typically recommended because it will eventually decompose and will add humus to the soil. This in turn improves the soil composition and texture. In addition, during decomposition nutrients are released, which increases the fertility of the soil. In selecting mulches, there are a number of factors to consider: the availability of the material, the cost, and the appearance. Local nurseries, or an extension service, should be able to advise you on the type of mulch suitable for your needs.

Prior to mulching, all beds and tree rings should be defined and edged. The edge should be maintained throughout the season to give the landscape a clean and crisp appearance. Edging debris should not be placed in the beds or rings but should be removed from the site since excess soil at the base of plants can be detrimental to their health. All tree rings should be evenly concentric around the tree and all bed edges should be maintained as one smooth and continuous line.

The beds and tree rings should be mulched to a depth of 2″ to 3″ in early spring. Any mulch existing from previous years that is in excess of 2″ deep should be removed or worked into the soil before new mulch is applied. Often the mulch is left in the beds and after years it builds up to 8″ to 10″. This is not a case where more is better. Since roots need oxygen to survive, excess mulch will prevent oxygen from penetrating the soil and the plants can suffocate. Excess mulch will also cause roots to grow into the mulch and not into the soil. This causes the plants to be shallow-rooted, which in turn causes the plants to dry out quickly and suffer during droughts.

It is a common practice in many areas to mound mulch around the trunk of a tree. This practice should be avoided for a number of reasons. First, mulch, if kept in contact with the bark, can promote attack by insects and disease. Second, the mulch will keep excess moisture on the trunk or crown of the plantings, which can cause decay.

In midsummer, the mulch should be lightly raked and loosened to break up any water-impermeable layers. A light top-dress application of mulch should be applied in early fall and at any other time during the season to maintain a consistent 2″ layer of mulch in all beds and tree saucers.

All beds, tree rings, and planting areas should be kept weed free at all times. Weeds should be controlled by hand. A pre-emergent herbicide may be applied before mulching to reduce weed germination. Post-emergent, nonselective contact herbicides should only be used as spot treatments.

### Pruning

There are many reasons to prune: to keep plants healthy, to restrict or promote growth, to encourage bloom, to repair damage, to remove structurally weak or otherwise undesirable branches, to clear a building, or to allow light to penetrate to the ground. Trees, shrubs, and ground covers all may require some type of pruning during the growing season to achieve these goals.

### Shrubs

It is important to remember that different types of shrubs have different growth habits and characteristics. Plants are selected for a particular area based on form, color, and texture. If all the plants on a site are sheared into hedges, or as individual balls, or squares, the characteristics of the plants are lost. Therefore, it is important to follow proper pruning techniques so that the natural beauty of a plant is recognized.

The best time to prune shrubs depends on their flowering habits. Shrubs which flower on new growth should be pruned in early spring before new growth emerges or during the last weeks of winter. Shrubs which flower on old growth should be pruned directly after flowering. If these shrubs are pruned throughout the growing season, the flower buds will be removed and there will be no flower display. As a general rule, spring-flowering shrubs should be pruned immediately after blooming. Broadleaf evergreen trees and shrubs should be pruned after new growth hardens except for hollies, which should be tip-pruned in early spring. Conifers should be pruned by pruning new growth (candling) and again, only if necessary, after the new growth hardens off. Shrubs that flower in summer should be pruned in late fall to early winter or early spring.

If plants are healthy, pruning should occur to maintain the shape of the plant and to encourage new growth. One-fifth to one-fourth of the old branches should be removed from the ground. This will encourage new growth from the base of the plant and will maintain the natural shape of the shrub. If additional pruning is necessary, hand pruning is appropriate as long as the natural habit of the plant is maintained. Proper pruning cuts are essential to maintaining the health of the plant (see Fig. 7.2).

Plants that are disease- and insect ridden and unmanageable require rejuvenation pruning. The first step in rejuvenation pruning is to remove old and diseased wood. The second step is to cut back healthy wood to encourage branching. Lastly, the sucker growth should be thinned. The key to rejuvenation pruning is to strive for a well-balanced and uniform plant.

**Figure 7.2** Proper pruning cuts. (1) A proper pruning cut is achieved by making a clean 45° angle cut away from the bud. The cut should be made about $^{1}/_{4}$" above the bud. (2) If a cut is made too close to the bud the bud may die. (3) A long stub will be unsightly and will eventually decay.

There are occasions where a plant, or group of plants, is completely overgrown or scraggly. Then it is appropriate to cut the plant to the ground in the early spring and allow new growth to renew the plant. Although this may seem extreme, certain plants will come back healthy and strong within one to two seasons. Check with your local nursery to see which plants will survive renewal pruning. If a plant cannot be cut to the ground, pruning can occur over a three-year period, each year cutting back one-third of the oldest canes. Whether a plant is cut to the ground or cut back a third at a time, the plant should be fertilized and watered to encourage new growth.

Hedges should be pruned by hand as necessary to maintain a neat and trim appearance. Hedges should be maintained at an exact and equal height for the entire length of the hedge and should be shaped with the bottom of the hedge slightly wider than the top (see Fig. 7.3). Rejuvenation pruning of old hedges, shrubs, and ornamentals is desirable.

### Trees

Trees may require pruning, particularly if they have been neglected for many years. All dead, diseased, weak, and cross branches should be removed to improve the structural integrity of the tree. To avoid having to extensively prune large trees, the trees should be pruned and trained while young. Properly pruned trees will grow into structurally sound trees as they mature. Cross branches should be removed, permanent branches should be carefully selected, and a strong branch structure should be developed (see Fig. 7.4).

Typically, the contractor should be responsible for all pruning that can be reached from the ground (or the bed of a truck on all roadways and parking lots) with an extended pole pruner. If a tree requires climbing into the tree, it should be pruned by a certified arborist.

Any trees or shrubs that are pruned to the point where the aesthetic quality of the plants is severely damaged or the health of the plants is jeopardized should be replaced entirely at the contractor's expense.

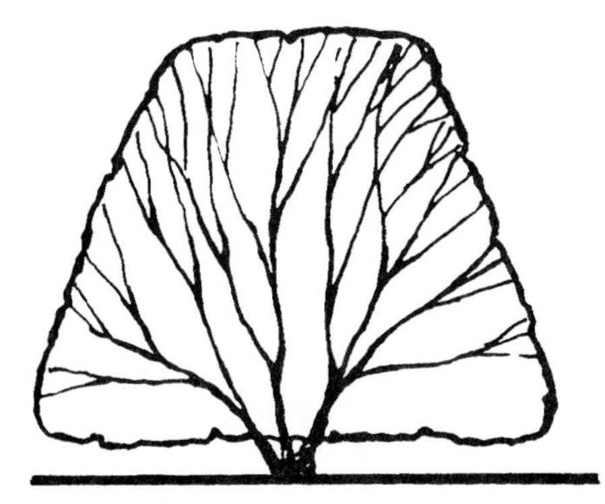

**Figure 7.3** Hedge pruning. Proper pruning for a hedge leaves the bottom of the hedge wider than the top. This allows sunlight to all areas of the shrub.

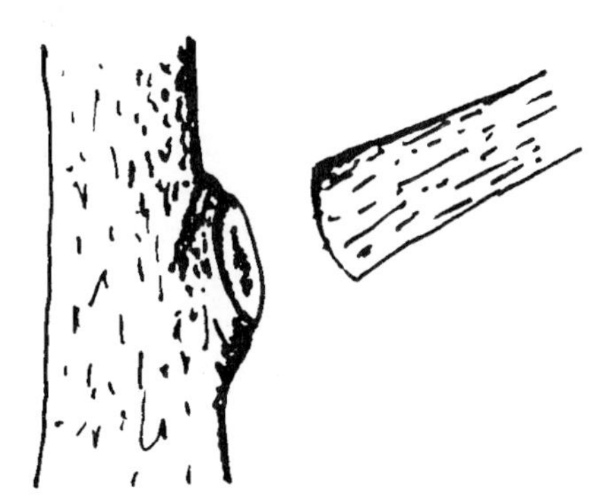

**Figure 7.4** Tree pruning. Proper pruning of a branch is cutting as close to the branch collar as possible. The branch collar should not be removed.

### Vines and ground covers

Vines and ground covers should be pruned regularly to maintain a neat and manicured appearance but should not be sheared. String trimmers are never to be used to prune ivy or other ground covers. Ground covers should be pruned at the nodes, with the cut hidden. Depending on aesthetic preferences, ground covers should be maintained within or partially overhanging all planters and off all paved surfaces.

Ground covers should be maintained 4″ to 6″ away from the trunks of trees and shrubs.

### Fertilization

Plants need different amounts and proportions of nutrients to stay healthy. In most areas, even if the soil has ample amounts of organic matter, supplemental fertilization will be necessary. Many soils have insufficient amounts of one essential nutrient with an overabundance of another. As discussed earlier, the amount and proportions of fertilizer required should be determined by a soil test.

Assuming that fertilization is required, shrubs and ground covers should be fertilized in early spring. Ericaceous (acid-loving plants) plant material should be fertilized with an acidifying fertilizer. Deciduous and evergreen trees should be fertilized in late fall after leaves have dropped and the danger of forced growth is past.

All trees and shrubs indicating chlorosis (a systemic condition in which new and old growth turns yellow) should be fertilized with chelated iron and other micronutrients as needed. Root-stimulating hormones should be applied to all plants that are in poor condition. Soil drenching is preferred over foliar application to prevent phytotoxicity (burning of the leaves).

### Cleanup

All areas, including planting areas, plant materials, lawns, and paved areas, should be kept clean at all times. The contractor should remove and dispose of any and all trash (including cigarette butts), sticks, natural debris (including soil, sand, rocks and gravel, withered flower buds, seed pods, leaves) from all landscaped areas, including all raised planters, turf, and ground cover beds during each visit.

In autumn, leaves should be raked and removed on a regular basis. All leaves should be removed from all lawn and bed areas before mowing, including leaves and branches that drop throughout the spring and summer months.

### Pest management

The contractor should be responsible for the detection, monitoring, and control of all pests. The contractor should be aware of the potential pests that might be encountered and should make regular and thorough inspections of all plant material. Treatment should be applied as necessary using products and methods that target the insect pest with minimal residual effects.

It would be difficult to discuss all the potential pests which can affect plantings. What is important to understand is that healthy plants perform well. How do you keep your plants healthy? Locate plants in a location that is suitable for the particular plant, prepare the soil well, water appropriately, and fertilize when necessary. All this is part of the philosophy of integrated pest management (IPM). If plants are in a healthy environment they are less likely to have diseases and pests. If they do have a particular problem only the problem area should be addressed.

The practice of spot treatment greatly reduces unnecessary pesticide use. It should never be assumed that if there is a problem in one area it must be widespread and therefore everything should be treated. This is often a waste of time and money. It is also important to realize that a low population of pests does not necessarily mean that chemical control is required. Since no landscape can be kept pest free, a management program should be developed that maintains pest populations below a damaging level.

If there are beneficial insects (ones that feed on the pests), then no chemicals need to be applied. Applied IPM techniques minimize the need for pesticides to control problems as well as decrease the probability of problems in the first place. If the philosophy of IPM is used, there should be less need for chemical applications, which ultimately lowers the cost for maintenance while enhancing the performance of the landscaping materials.

If there is a need for chemical application, the contractor must adhere to the Department of Agriculture Regulations for Commercial Application of Pesticides. All pesticide applicators should be licensed or directly supervised by a licensed applicator. All licenses should be for commercial application and should be current.

The contractor should inform the facility manager of the pesticides to be used on the property and should receive approval from the facility manager before making substitutions on specified chemicals. The contractor should submit copies of product safety sheets of all pesticides used to the facility manager to be filed in the departmental office.

Upon completion of each pesticide application, the applicator should record all information on a data sheet and file it in the facility department office (see Fig. 7.5).

Pesticide applications should not be done as calendar-scheduled or general cover sprays. Pesticides should only be applied as needed, when pests are detected through regular inspections. Pesticides should be applied at a time of day when human activity is at a minimum in the facility.

### Watering

The key to watering is to water deeply and infrequently. This helps to develop an extensive, deep root system. Frequent, light waterings encourage roots to stay near the surface. This encourages the plants to be more and more shallow rooted. Shallow-rooted plants tend to be less vigorous and suffer during drought. It is always preferable to water early in the morning so the sun will burn off the excess moisture. This will decrease the potential for fungus and disease to set in.

If plants are too dry their leaves and flowers will wilt and the plant will eventually wither and die. Plants can also die from too much water, especially if the water accumulates around the roots of the plant. With too much water, the leaves and flowers turn black and fall off and the roots will rot. When the roots are exposed, they will be black with no healthy white roots. The exposed roots will often smell foul. The correct amount of water will vary according to soil type, plant and turf species, climate, and weather. Once the correct amount of water is determined, plants will be healthy and vigorous.

PESTICIDE APPLICATION RECORD AND DATA SHEET

Applicator's Name(s): ____________________

Date and Time of Application: ____________ 19____ . ________ M.

Weather Conditions (Temperature, Wind Direction and Velocity, Precipitation, etc.): ____________________

____________________

Pesticide Used - Label Name and Formulation: ____________________

____________________

Concentration: ____________________

Total Amount of Pesticide Mixed: ________ Amount of Pesticide Used: ________

Amount of Pesticide Left Over: ____________

Disposal of Leftover Pesticide: ____________

Application Equipment Used: ____________________

____________________

Safety Equipment Used: Face Shield ( ) Goggles ( ) Respirator ( ) Gloves ( )
Boots ( ) Headgear ( ) Tyvek Suit ( ) Rainsuit ( )
Other ____________________

Plants Treated: ____________________

____________________

Number of Plants or Area Treated: ____________________

____________________

Reason for Pesticide Application: ____________________

____________________

Applicator's Comments: ____________________

____________________

____________________

____________________

____________________
(Applicator's Signature)

**Figure 7.5** Sample pesticide application and data sheet.

It is best to have one person responsible for the proper and adequate irrigation of all lawn areas, beds, trees, annual flowers, and all other plantings throughout the season. In most situations, watering should be the full responsibility of the landscape contractor, since the contractor is most familiar with the plant needs.

Automatic underground irrigation systems should be used where they are installed and working. Again, the contractor should be solely responsible to coordinate the timing and control of these systems. Irrigation systems should be checked regularly to make sure heads are spraying correctly and lines have not been broken.

Areas without irrigation systems should be watered by hand. If the contractor is responsible for the watering, he should supply hoses, water trucks, and all other equipment needed to properly and adequately irrigate by hand.

### Seasonal color

Seasonal color is essential for distinguishing one property from the next. For very little money, annuals, perennials, and bulbs can enhance the overall appearance of a property.

Annuals are defined as plants that complete their life cycle in one season. In most areas of the country they will flower throughout the growing season, but will need to be removed at the end of the season. With careful planning and design, the display will perform for the entire season.

Perennials are plants that will come back each year, but will typically have a shorter blooming time. Mixing perennials which bloom at different times will give a constant display of color. Since perennials will come back year after year, it is recommended that a landscape architect, or garden designer, be involved in the layout and design to ensure a successful planting.

Bulbs have a short blooming period but give a beautiful display and are always welcome in the early spring. Many bulbs, such as tulips, are removed after they bloom. Others, such as daffodils, can be left in place for many years as long as the bed is not disturbed.

The seasonal color displays should be unified rather than disjointed groupings throughout the property. To achieve the greatest visual impact, the displays should be planted at key points to a building such as entrance drives and the front door. Keep the designs simple, using a few colors and a limited number of plant types. A single color will have more impact than five or six colors mixed together.

Annual planting beds are high maintenance, and this should be considered in determining the size of the bed and the location. Annual plantings may need daily watering and will need regular fertilization. Scattered beds throughout a property that do not have access to water will decline quickly and show poorly. On the other hand, a small bed near the front entrance will be noticed by everyone entering a building and will be easier to maintain.

Soil preparation is one of the most important elements in achieving a successful seasonal display. Seasonal plantings need good drainage and fertile soil to look their best. First, the soil should be tested for pH to determine whether it is acid (6.9 and lower) or alkaline (7.1 and higher) and if there are any nutrient deficiencies. For the most accurate results, a soil sample can be sent to a soil testing laboratory.

Next the area should be checked for drainage. An easy test to assess whether there is proper drainage is to dig a hole 1 ft × 1 ft × 1 ft and fill it with water. If the water percolates out within 1 to 2 hours the drainage is adequate. If it takes more than 1 to 2 hours, some type of drainage system may be necessary such as drain tiles or a dry well. It is important to note that very few plants can survive in wet conditions, so poor drainage should always be corrected.

Once the drainage is checked and the area is draining adequately, the soil should be amended. Amending soil with organic material will improve the soil quality and improve aeration and may also improve drainage. As the organic matter decomposes, it will add nutrients into the soil which then can be taken up by the plants.

The plant bed should be tilled to a depth of approximately 12″. The additives should be worked into the soil, and the consistency of the soil should be light and airy. Once this is completed, the seasonal display can be installed.

## Quality control

Quality control should be the responsibility of the landscape contractor. The landscape contractor should employ site supervisors and grounds technicians who accept responsibility for the appearance of the site. Inspections of the property should occur on a regular basis (monthly or quarterly) to establish crew goals for quality improvements as well as to identify the site's needs. Copies of the inspection reports should be sent to the facility manager for review.

**Landscape Maintenance Schedule**

| TASK | Jan | Feb | Mar | Apr | May | Jun | Jul | Aug | Sept | Oct | Nov | Dec |
|---|---|---|---|---|---|---|---|---|---|---|---|---|
| Spring Clean-up | | | | ✓ | | | | | | | | |
| Mulching | | | | ✓ | | | | | | | | |
| Mowing | | | ✓ | ✓ | ✓ | ✓ | ✓ | ✓ | ✓ | ✓ | ✓ | |
| Remove Litter/Debris in Turf Areas | | | ✓ | ✓ | ✓ | ✓ | ✓ | ✓ | ✓ | ✓ | ✓ | |
| Weed Control | | | ✓ | ✓ | ✓ | ✓ | ✓ | ✓ | ✓ | ✓ | ✓ | |
| Pre-emergent Herbicide | | | | ✓ | | | | | | | | |
| Grub Control | | | | | ✓ | ✓ | ✓ | ✓ | ✓ | ✓ | ✓ | |
| Post-emergent Herbicide | | | | ✓ | | | | | ✓ | | | |
| Fertilize Shrubs/Trees | | | ✓ | | | | | | | | | |
| Fertilize Turf | | | ✓ | | | | | | | | ✓ | |
| Core Aerate Lawn | | | | | | | | | ✓ | | | |
| Dethatch Lawn | | | | | | | | | ✓ | | | |
| Soil Tests | | | | | | | | | ✓ | | | |
| Over Seed | | | | | | | | | ✓ | | | |
| Lime | | | | | | | | | | | ✓ | |
| Dormant Oil | | | ✓ | | | | | | | | | |
| Other Pests | | | ✓ | ✓ | ✓ | ✓ | ✓ | ✓ | ✓ | ✓ | ✓ | |
| Pruning Trees | ✓ | ✓ | | | | | | | | | | ✓ |
| Pruning Shrubs/Ground Covers | | | | ✓ | ✓ | ✓ | | ✓ | ✓ | ✓ | | |

**Figure 7.6** Typical landscape maintenance schedule.

The landscape contractor should familiarize all site personnel with the specific requirements of the site and the specifications for maintenance.

The facility manager should also perform quality control inspections with the contractor. Figure 7.6 is a landscape maintenance schedule that can be used as a checklist to ensure that all tasks are being performed and that all deficiencies are being addressed and actions are taken to rectify them. The schedule should be modified to meet the specific requirements of the site and the specifications.

# NOTES

# NOTES

Chapter

# 8

# Elevator and Escalator Services

## Operating Instructions

Signs that are required by ASME A17.1-1996 Safety Code for Elevators and Escalators are listed in this chapter. It should be noted that your local code authorities may have adopted only a portion of this national code or may have made their own revisions. Currently the A17.1 code is being combined or harmonized with the B44 Canadian code and will change many rules when the harmonization efforts become the standard. Also the version of the code in effect at the time your equipment was installed or modernized governs the rules you must follow except for retroactive rules. Symbols for blind or sighted passengers (American Disability Act–ADAAG) and signage required for firefighter's operations are not shown below, as vintage and local codes dictate the exact requirements. Your local code-enforcing authorities can help with any questions.

## Elevators

Blind hoistway and emergency doors must have the sign (at least 2″ lettering) *danger, elevator hoistway* posted on the corridor side of the access door.

A photograph as shown in Fig. 8.1 should be posted over each elevator corridor call station. It is 5″ wide and 8″ high.

A plate stating the capacity (pounds and number of passengers) and the rated load should be fastened in a conspicuous place inside the elevator. Lettering should be 14″ or larger.

## Freight elevators

Signs (letters at least ½″ high), in addition to the capacity and data plates that should be permanently and conspicuously posted inside the cab, should include:

- For class A freight elevators permitted to carry passengers: *Class A loading. This elevator designed for general freight loading*
- For class B freight elevators permitted to carry passengers: *Class B loading. This elevator designed to transport motor vehicles having a maximum gross weight not to exceed lb*
- For class C-1 freight elevators permitted to carry passengers: *Class C-1 loading. This elevator designed to transport loaded industrial truck maximum combined weight of industrial truck and load not to exceed lb*
- For class C-2 freight elevators permitted to carry passengers: *This elevator designed for loading and unloading by industrial truck maximum loading and unloading weight while parked not to exceed lb. Maximum weight transported not to exceed lb*
- For class C-3 freight elevators permitted to carry passengers must meet the requirements for passenger elevators: *Class C-3 loading. This elevator designed to transport concentrated loads not to exceed lb*
- For the above freight elevators that are *not* permitted to carry passengers, the sign should read: *This is not a*

**Figure 8.1** Elevator corridor call station pictograph. (*Courtesy of ASME.*)

*passenger elevator. No persons other than the operator and freight handlers are permitted to ride on this elevator.*

### Hand elevators

Hoistway doors must have the following sign (at least 2″ lettering) posted on the corridor side of each door: *Danger-elevator-keep closed*

### Hand-operated dumbwaiters

Each hoistway door should have the following sign posted on the corridor side with letters not less than 2″ high: *Danger-dumbwaiter-keep closed.*

### Hand- and power-operated dumbwaiters

A sign stating *no riders* should be located in the car in letters not less than $^1/_2$″ high.

### Wheelchair lifts

A passenger restriction sign should be provided at each landing and on the platform. It should be securely fastened in a conspicuous place and state *physically disabled persons only—no freight.* The sign letters should not be less than 14″ high and should include the international symbol for physically disabled persons.

### Escalators and moving walks

A caution sign shown in Fig. 8.2 should be located at the ends of the escalator or walk visible to the boarding passengers. The size should be 4″ wide and $7^3/_4$″ high. The ends should contain a red-colored stop button containing the words *emergency stop.*

### ADA signage (American Disabilities Act)

Signage is required to aid the handicapped. While most of the regulations are aimed at Braille operating controls placed at wheelchair-accessible levels in the elevator, some signage may be required.

### Additional signage

Caution, warning, or traffic regulation signs may be placed in areas where appropriate to deter equipment misuse or improper operation. While only the signs stated above are required by code, these additional signs can be helpful. An example would be a sign or caution tape placed across an elevator entrance reading *Elevator being serviced.* Another example would be a sign on the equipment room door reading *Danger, elevator equipment.* Signs on switches and disconnects controlling equipment lighting and power circuits reading *Always leave on* may help to prevent inadvertent shutdowns or entrapments.

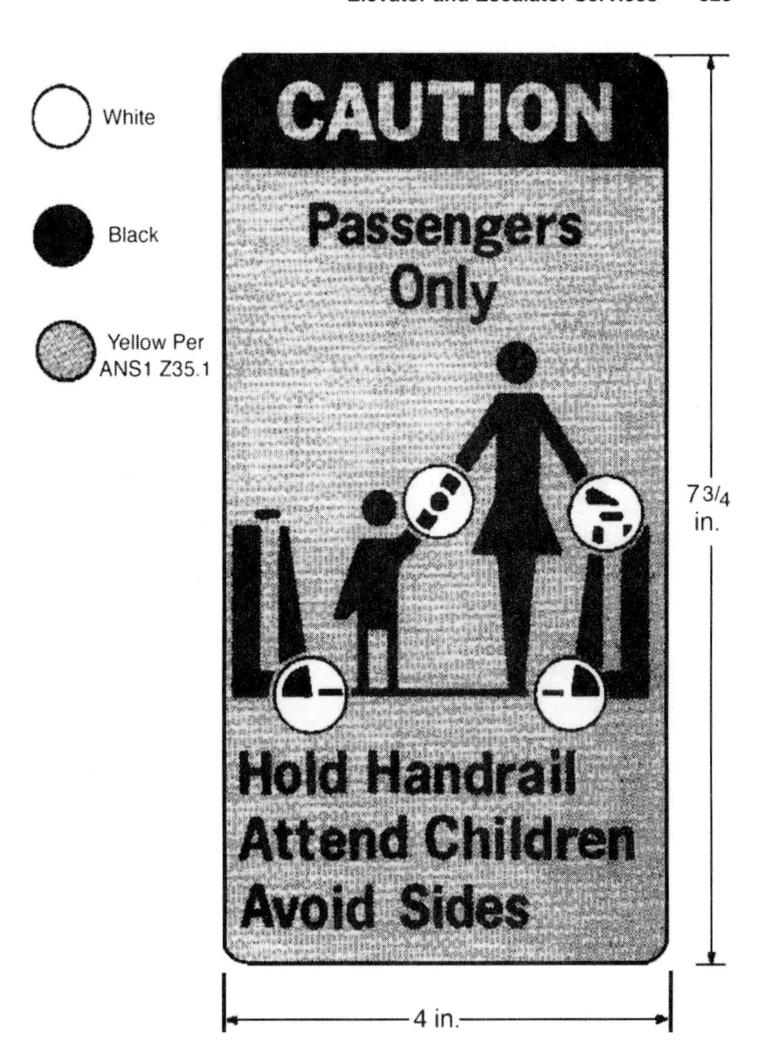


**Figure 8.2** Escalator/moving walk sign. (*Courtesy of ASME.*)

### Equipment identification signs

All vertical transportation equipment within the building should be marked with a unique identification number. Where practical, mark each elevator sequentially from 1 through 24 if there are 24 elevators in the building. All components of the elevator should contain the same number on the control panels, the machine, motor generator, the governor, the main line power disconnect, the pit equipment, and inside the elevator. This will help ensure that a component is not mistakenly serviced, resulting in personal injury or equipment damage. It also permits passengers to report a faulty operation of an elevator or correctly report an entrapment. Floor markings should be placed on the leading edge(s) of the hoistway doors, visible when the door(s) open so that a problem at a specific floor can also be communicated to the repair technician or can identify a floor to emergency personnel. Escalators, dumbwaiters, and other unique equipment can be identified with similar markings of elevators, since there would be little chance of confusion.

Operating instructions and signage described above will help ensure proper use of the equipment. It does *not* provide the passengers with the best time of day to use the equipment. The design speed and capacity of the equipment were based on the projected traffic flow from 7 A.M. to 7 P.M. on normal working days. The peak periods of passenger usage for most office buildings will occur between 7:30 A.M. and 9 A.M. for up-traveling passengers and between 4:30 P.M. and 6 P.M. for passengers leaving the building. Lunch time peaks usually occur between 11:30 A.M. and 1 P.M. The number of units required to prevent "bunching" of passengers (e.g., 30 or more people waiting at the lobby floor to board an elevator at 8 A.M.) in the building lobby and other corridors was considered by the designer. For more information on the application of the units see *Vertical Transportation: Elevators and Escalators,* 2d ed., by George Strakosch or *Vertical Transportation Standards,* 7th ed., by the National Elevator Industry, Inc. Both can be obtained through the Educational Materials Department of *Elevator World Magazine.* To minimize bunching and other long corridor waiting times in your building, and therefore optimize equipment performance, consider one or more of the following actions:

- Keep all available equipment "in service" during peak periods, avoiding planned maintenance shutdowns, deliveries, and other utilization of the equipment that could be postponed to in between or after peak periods
- Keep the elevator supervisory (dispatching) systems in proper working condition and their time clocks set to the correct time of day
- Run pairs of escalators in the directions favoring the flow of traffic
- Where practical, implement staggered starting times and lunch periods for building personnel
- Encourage building personnel to follow good passenger practices, for example, no registering of both corridor push buttons in an effort to get an elevator more quickly
- Discourage passenger flow opposite to the major traffic flow where possible

Contact your elevator contractor or consultant if you encounter problems that may only be corrected by the addition of special features.

## Maintaining Equipment

### Definitions and terminology

**Traction elevator.** A traction elevator is a cab and counterweight system (see Fig. 8.3) connected by a set of several steel ropes. The ropes are routed over the sheave (grooved wheel) of a driving machine. The machine motor can drive the sheave directly (*gearless traction elevator*) or indirectly with worm and gears (*geared traction elevator*). Owing to design requirements, the traction elevator requires more maintenance time when compared to all other types of vertical transportation equipment in the building.

**Hydraulic elevator.** The hydraulic elevator is a cab connected to a jack consisting of a plunger and cylinder (see Fig. 8.4). The most common types of hydraulic elevators have

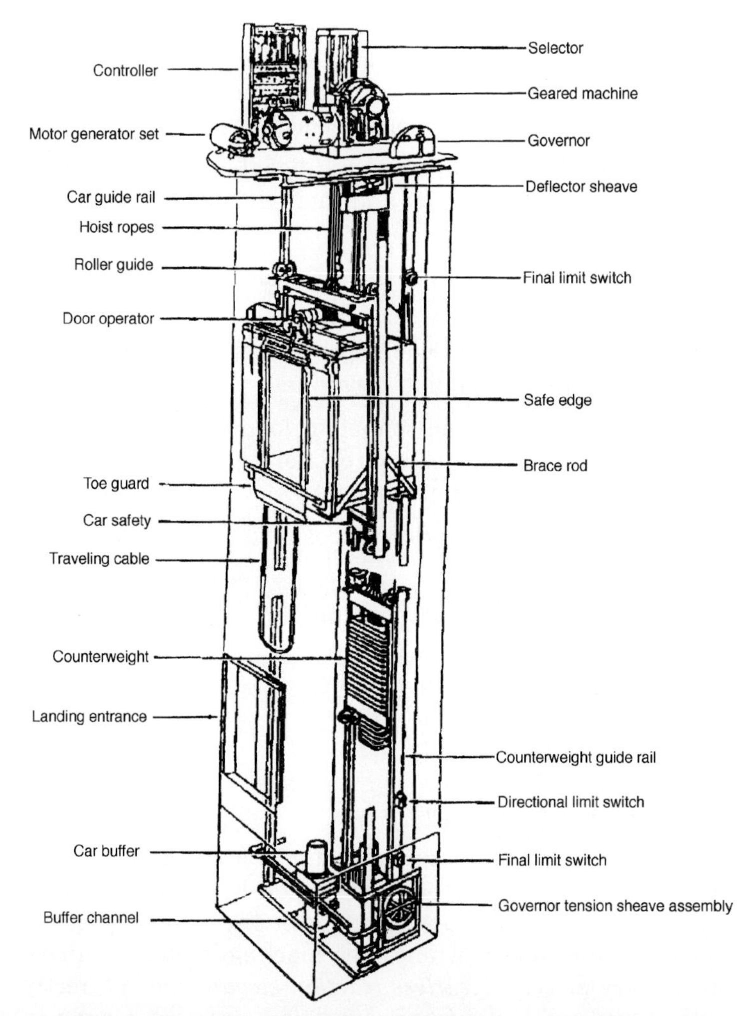


**Figure 8.3** Traction elevator. (*Courtesy of NEIEP, Elevator World.*)

underground jacks with plungers that push the bottom of the cab upward to the floors (eight or fewer). The jacks reside in the ground to a depth approximately the same as the travel of the cab. An electric motor and pump, using hydraulic oil, propel the plunger and cab upward. The motor and pump are idle in the down direction. Gravity causes the

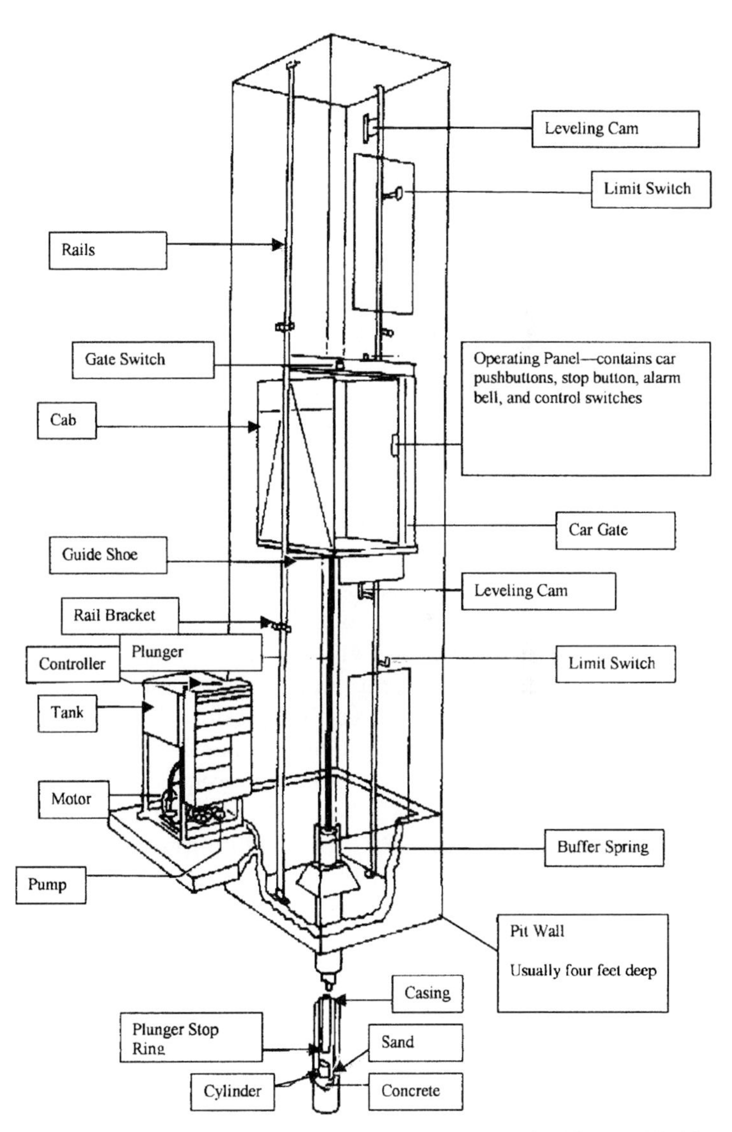


**Figure 8.4** Hydraulic elevator. (*Courtesy of Otis Elevator Co., Elevator World.*)

cab and plunger to descend at a speed controlled by a valve in the oil line. The conventional hydraulic elevator is the least complex type of passenger and freight elevator and will require two to four times less maintenance than the traction elevator. Maintenance costs should reflect this ratio.

**Maintenance tasks.** Tasks are the actions performed on the equipment. There are six actions—examining and testing are the first two actions to determine which, if any, subsequent actions are required. These are cleaning, lubricating, adjusting and repairing, or replacing a part or all of a component (or cleaning equipment areas such as the car top, pit, or machine room). The tasks are necessary to help maintain equipment in a safe and efficient operating condition.

**Routine maintenance examinations.** A set of observations and tasks is performed each time the unit is scheduled for maintenance. The typical routine examination includes equipment ride quality with an emphasis on leveling, door operation, and the signal lights and bells. A machine room check is made of all rotating equipment for oil levels and leaks, brushes and commutators, brake operation, and relay functionality. Escalators and walks will require examinations of the steps, comb teeth, handrails, key switches, and brake stopping distances. Routine examination frequency on elevators and escalators will vary owing to age, condition, operating environment, and usage. More maintenance time is required for older vintages (with the same condition and usage) unless the equipment has been modernized with state-of-the-art components.

**Periodic maintenance tasks.** The tasks include equipment tests performed quarterly (every 3 months) or at different intervals. Most equipment tests are done periodically. The typical periodic examination is made in three general areas: (1) machinery room(s), (2) car tops, and (3) pits on elevators; the upper and lower pits and the incline on escalators and walks. Brakes, moving selectors, controllers, governors, and tests will require periodic maintenance one to four times annually. Cleaning, car shoes, rope examinations, doors, door operators, and locks will be included in the car top area examination and will be made one to four times each year. From the pit, the lower car equipment, the buffers, compensation, and governor tension sheaves must be inspected, and this is usually done every 3 months.

**Maintenance schedule.** This is a list or chart that contains routine and periodic tasks, as well as the interval between the tasks. The intervals can be based on calendar days or on usage of the equipment, e.g., number of stops on an elevator or the hours of operation on an escalator. The schedule directs the technician as to *what* tasks to perform and *when* to perform them. When the tasks are completed, the schedule can be referred to as the maintenance log. The schedule and log can be either a written document that resides at the job site or an electronically transmitted database from the equipment's control processor, stored on a computer.

**Maintenance guide.** This is a procedure or a group (book, document, or database) of procedures describing in detail *how* a task, exam, or test is to be performed.

**Time-based maintenance.** A fixed amount of time between scheduled maintenance. For example, a traction elevator may receive *scheduled* maintenance monthly, a hydraulic elevator every 6 weeks, and an escalator every 8 weeks.

**Use-based maintenance.** These types of scheduled examinations and periodic maintenance tasks are dependent on an event—for example, the number of stops or miles traveled on an elevator, or the number of operating hours on an escalator or walk. For example, 8000 stops could schedule a traction elevator routine examination; 4000 stops may schedule an examination on hydraulic elevators; and 600 hours may schedule an escalator or walk (depending on vintage, operating environment, and condition of equipment).

Microprocessor-based equipment has features that facilitate the acquisition of this information, and in some cases the information can be remotely accessed.

**Use- and time-based maintenance.** This frequency of maintenance scheduling is a combination of the use-based and time-based maintenance scheduling defined above. Past history may indicate that the elevator, for example, makes 8000 stops in a six-week period. Then for the next 12 months, the routine examinations would be performed at six-week intervals. If the number of shutdowns (callbacks) and customer satisfaction inquiries indicate that the interval is adequate, then it may be continued at that frequency. This method of scheduling proves to be more effective on older equipment where stop counts and running hours are not readily available. Where stop counters and time accumulators are installed on the units, the routine exams can be made at fixed intervals of time and the periodic exams can be made based on usage data. This approach is productive when more frequent visits are required.

**Interval.** This is the period of time between examinations. A fixed interval implies *time-based* approaches to maintenance. A *use-based* maintenance approach would generally imply a variable or floating interval.

**Callback.** This is the term that is used for an *unscheduled* visit by the service technician to an elevator or other type of vertical transportation unit requiring some form of service. Callbacks may occur owing to an equipment mis-operation, misuse or abuse, a need for transporting large loads on the elevator under supervision by the elevator technician, or recovery of a lost item inside the elevator hoistway or escalator.

**Availability.** This indicates the amount of time that the equipment is available to the user, usually indicated in hours or percent. An elevator normally available for passenger use may be down for 34 hours in 1 month, and therefore available only 80% of the time. Causes for downtime include equipment maintenance, breakdowns, and planned shutdowns ("mothballing") due to passenger inactivity in the building.

**Repair work.** Any restoring action taking more than 4 hours and/or needing two or more technicians.

**Openings.** The floors where passengers or freight can ingress or egress. There may be one or more openings at each floor that the elevator serves.

### Contracted maintenance

This is an agreement between the facility and a maintenance contractor. Contracts include maintaining most equipment components in a safe and efficient operating condition. Some components are excluded, as shown in the "sample contract." Contracts are usually written to cover a period of 1 to 5 years with a renewal clause after the initial term. Cancellation can occur when a party to the agreement gives 30-day notice of a desire to terminate the agreement. While some buildings hire personnel to maintain their equipment, contracted maintenance, originating in the early 1920s remains as most popular.

Companies can be chosen from the firm that originally installed or upgraded (modernized) the equipment or from a company that only performs maintenance and not installation and modernization. At this writing, several U.S. and most European companies are ISO 9000 certified. This or an equivalent certification requires a rigorous documentation of operating procedures by the company as well as periodic compliance inspections by the certifying organization. While many companies may do good maintenance, any need for engineering and extensive modifications can usually be done more efficiently by the original equipment manufacturer, having access to the original design and application database. Examples would be modifications of speed and/or capacity, or adding an additional unit. Before deciding which company will be awarded the contract, ask the contractors to show you both their maintenance-operating plan and their plan for upgrades, as it would apply to your building. When done correctly, this plan describes *how* the company will execute their contractual requirements. Review the plan carefully with your elevator consultant.

If the equipment at your facility has just been installed then traditionally you will receive at least 90 days of new installation service by the installing company. This is done to make further adjustments to the equipment as it gets broken in. At the end of the free service period the installing company and other maintenance companies will as a rule call on you offering a service contract. From the installing company it will be very important that you obtain a complete set of "as built" external wiring diagrams. The set can vary from several to one hundred or more sheets of paper. Without the diagrams, no company can provide proper service to the equipment. Keep the diagrams with your important records and provide a copy to the maintenance company if it is different from that of the installing company. It will be necessary to have these drawings updated when any major changes are made. Keep one copy as a master to reflect these changes. Examine carefully each company's contract for the following points:

- Are the major components covered in full, with parts and labor included if replacement is required? The hoist motor, brakes, motor-generator set (if it exists), all ropes, cab or escalator and walk structure, controllers, governors, selectors, guide shoes, and rails for both the car and counterweights should be included. Some contracts include only oil and grease (lubrication) and charge for all parts needed, while some supply only normal renewal parts such as contacts and motor brushes along with the lubrication.
- Most contracts will *exclude* the following items: cabs, cab finishing and flooring; all doors and door frames and sills; light fixtures and lamps; cover plates for fixtures and operating stations; power switches and breaker feeds to

the equipment; smoke detectors; external emergency power equipment feeds; underground piping and cylinders on hydraulic elevators; and the renewal or refinishing of all exposed trim on escalators. This is a fair practice, since most of these items are beyond the elevator maintenance company's control.

- While most maintenance contracts offer only overtime callbacks, most repairs are to be done during regular working hours. For a premium, planned repairs on equipment critical to the building are done on overtime or building holidays.
- Insurance and governmental tests should also be included. Some tests, however, can best be done by building personnel during off-hours and should not be included in the contract. These tests include a monthly check of the firefighter's service, emergency power tests, and tests of life support systems. Doing these tests during off-hours can be more productive and much less disruptive to the traffic flow in the building. Smoke detectors should be tested by specialists as prescribed in NFPA-72.
- Hoistway cleaning on elevators and truss oil pan cleaning on escalators should be included to avert a several hundred dollar annual cost.
- Overtime callbacks should be included. These are callbacks other than 8 A.M. to 4:30 P.M. Monday through Friday. A typical contractor will pay the straight time costs, with the building incurring the bonus labor for overtime work.
- Exclusions will almost always include consequential damages caused by the equipment being unavailable for use. Equipment obsolescence creating unavailability of parts and other material will usually be excluded. Have your consultant review this carefully, as it can be critical as equipment age exceeds 20 years.
- Performance bonds which may be in the form of a penalty for not meeting expected goals relating to the number of callbacks, equipment availability (see definitions above),

and corridor call waiting times could be included in the contracts. Have an elevator consultant guide you as to how the equipment is to perform, and/or to write the performance bond into the contract.

After the contract is awarded to a maintenance contractor, the following should be established:

- The day of the week or month routine examinations will be performed.
- The building person that will be screening any requests for call-backs (so the company does not respond to bogus calls).
- An in-house log-in, log-out book or system for the elevator technician.

This is the alternative to contracted maintenance. While the up-front costs of staffing, training, and equipping elevator personnel to maintain and repair the vertical transportation in the building can be large, it can have long-term paybacks. Some situations where in-house maintenance should be considered include remote areas where elevator service companies have no personnel; highly diversified and/or unique elevator equipment; high-security buildings where any outside personnel would require accompaniment by building personnel; or dissatisfaction with contracted maintenance performance in the area. Most in-house maintenance of elevator equipment is currently being done only in large complexes such as large real estate firms, universities, and industrial complexes.

Listed following are staffing guidelines to follow when considering in-house maintenance and repair of vertical transportation equipment. If the number of units in your building or complex is less than 40 traction elevators (20 each geared and gearless) or 70 mixed units (traction and hydraulic elevators, escalators, etc.) then consider having one maintenance technician. Contract out any major repair work needed with the local maintenance contractor, such as re-roping of elevators, large motor and bearing repairs, or any job taking more than 4 hours and/or requiring two people to re-rope. The guideline for the number of units required for two in-house technicians to perform *both* maintenance and repair would be 70 traction elevators or 175 equally mixed escalator and hydraulic elevator units. More units will require more people. Use these numbers as a guide for cost justifying in-house maintenance and the staffing required. However, it is important to keep in mind that the above loading is based on some assumptions. They are that the number of openings on the elevators are 15 or less, that the distance between work assignments is 10 minutes traveling time or less, that the equipment is operating in a clean and controlled environment, is in good condition, and is less than 20 years old. Reduce the technician(s) workloads for equipment falling short of these assumptions. If in doubt, consult with an elevator field productivity consultant for estimating your workloads, and by all means, monitor equipment performance periodically and adjust the workloads accordingly.

If you are currently performing in-house maintenance of the vertical transportation equipment, you should consider "maintenance alternatives." Included are "specialized maintenance assignments" and "amplified or dedicated maintenance." Depending on the size of your workforce, consider escalator specialists, hydraulic specialists, or certain traction elevator specialists, doing maintenance and/or repair functions on your vertical transportation equipment. A repair team may only perform major repair and/or adjusting of the equipment, a callback technician may only respond to callbacks, while the remaining technicians perform basic maintenance functions. Alternating the technicians periodically can broaden their expertise. The advantages in addition to the "economies of scale" are that the amount of specialized tools and transportation vehicles would be minimized, resulting in tool cost reduction; also the training time per technician would be minimized.

Equipment shutdowns or callbacks can occur throughout the normal workday or occur on nights and weekends. Each period, night or day, requires different administrative and management decisions and actions.

Elevator management people have perhaps their biggest challenge in orchestrating the workforce to answer overtime callbacks. Callbacks occurring on workdays between the hours of 6:30 A.M. and 8 A.M. and between the hours of 4:30 P.M. and 6 P.M. are the most difficult to orchestrate. Technicians should use mobile phones and pagers during these hours. Another method would be to utilize and pay overtime rates to standby personnel, if only to answer any callbacks involving entrapment of passengers. A technician standby list would be required. Take actions along these lines if callbacks are occurring during these hours. While costly, the callback response time can be reduced by 1 to 2 hours during these peak traffic hours.

The same rationale applies for all other "after hour callbacks." The service company develops a rotational list of technicians that would be called first to answer callbacks. If the person(s) on call is unable to respond to the callback or needs assistance, then other people not on the list would be called until someone is assigned to answer the callback. Lists with primary and secondary technicians (backup) will help. While this procedure may appear haphazard to some, the method is in wide use today.

Will there be enough elevator technicians to provide your large complexes with "after hour" callback coverage? Logistically it takes a technician 2½ to 3 hours to answer a single callback. Less time would be required for successive or back-to-back callbacks. Phones, pagers, and/or beepers are a must to schedule successive callbacks. Otherwise you may not have enough technicians to cover the area. Summer vacations and holidays will compound the problem. A log listing the history of your building(s) callbacks is necessary, not only for staffing purposes but also for liability defense if entrapment and/or injury occur.

A *maintenance control program* (*MCP*) or maintenance plan describes in detail *what* routine examinations, periodic maintenance tasks, and tests are performed by the maintenance technician, *when* (at what frequency) they are performed, and *how* they are to be performed. The MCP should describe the method used to document maintenance activity on each unit. This documentation is referred to as a maintenance log. The maintenance technician fills out the log while performing the maintenance activities. An example of a simple time-based maintenance log for an eight-car bank (group) of traction elevators is shown in Fig. 8.5. Use-based or combination use-based and time-based logs are more complex but can be more effective in logging maintenance activity. The MCP and the maintenance log should be customized for each bank of units. A maintenance guide describing how the tasks and tests are to be performed should also be included in the MCP. Building management, the elevator consultant, the maintenance contractor, or a combination of the three can develop the MCP.

### Removing equipment from service

Your watch engineer, or other authorized building personnel, will need to remove equipment from normal operation when malfunctions occur. The best place to turn equipment off is inside the elevator or at the top or bottom of the escalator, to avoid entrapments of passengers in the elevator or someone falling on the escalator. Turn the keyed lock in the elevator car station to the off position. Turn off both the "in service" and the "power switch." On some elevators the switches are behind the locked door in the car station. Turning off the cab lighting will deter people from entering the cab. Pull the doors closed if possible. For the escalator, simply make a visual check that no one is riding the escalator or about to board; then press the stop switch located at the top or bottom newel ends. A sign on the key switches will alert other personnel that a malfunction has occurred and to check with the building office prior to returning the escalator to service. The signs can also be placed inside and outside the elevator.

### Monitoring equipment performance

**What to monitor.** The following items need to be monitored as frequently as monthly, particularly on equipment built in

Building________________ Contract #________________

**Maintenance Log**

**Unit(s) #_____**

Technician______________ Start date______________ End date______________

| | Unit #___ | Unit #___ | Unit #___ | Unit #___ | Unit #___ | Unit #___ | Unit #___ | Unit #___ |
|---|---|---|---|---|---|---|---|---|
| **Routine #** | | | | | | | | |
| 1 | | | | | | | | |
| 2 | | | | | | | | |
| 3 | | | | | | | | |
| 4 | | | | | | | | |
| 5 | | | | | | | | |
| 6 | | | | | | | | |
| 7 | | | | | | | | |
| 8 | | | | | | | | |
| 9 | | | | | | | | |
| 10 | | | | | | | | |
| 11 | | | | | | | | |
| 12 | | | | | | | | |
| **Periodic Machine Room** | | | | | | | | |
| Lubrication | | | | | | | | |
| Controller | | | | | | | | |
| Brake overhaul | | | | | | | | |
| Blow out | | | | | | | | |
| Tests | | | | | | | | |
| **Hoistway** | | | | | | | | |
| Lubrication | | | | | | | | |
| Landing devices | | | | | | | | |
| Guides & rails | | | | | | | | |
| Ropes | | | | | | | | |
| Hoistway interlocks | | | | | | | | |
| **Pit** | | | | | | | | |
| Lubrication | | | | | | | | |
| Lower guides | | | | | | | | |
| Limits | | | | | | | | |
| Safties & buffers | | | | | | | | |
| Other | | | | | | | | |
| | Unit #___ | Unit #___ | Unit #___ | Unit #___ | Unit #___ | Unit #___ | Unit #___ | Unit #___ |

**Figure 8.5** Maintenance activity log. Pf, performance; Act, maintenance. (*Courtesy of SeeC, L.L.C.*)

the 1980s or earlier that has not had more recent control modifications. Later equipment has more reliable solid-state and computerized timing circuits; however, local code requirements still may require monthly testing of firefighter's service and the documentation of hydraulic oil additions. Your elevator consultant, elevator contractor, or local code-enforcing representative can provide you with the methods of testing required in your area.

- *Firefighter's service* (monitored monthly by building personnel. Test the system by recalling all elevators with the lobby switch-all elevators in normal service at that time should return to the lobby. Then test the firefighting operation of at least one of the elevators that returns to the lobby using the switch inside the elevator.
- *Emergency power* (monitored periodically by building personnel). Test the operation of the equipment that has been designated to operate on the emergency power supply system (diesel generator or secondary supply system) when normal power has failed.
- *Elevator performance* (monitored periodically by an elevator technician, witnessed occasionally by building management). Randomly select one or more elevators, or perhaps an elevator that you sense is running more slowly than the others, and have the elevator technician verify that it is performing within 10% of contractual requirements. Included will be the following: *contract speed* measured with a tachometer on the governor rope in the machine room (hydraulic elevators seldom have governors and the speed is measured from the top of the car; therefore, only the elevator technician, for safety reasons, should be checking these speeds); *door opening and closing times* measured with a digital stopwatch (times will vary depending on door system types; however, opening speed will exceed the closing speed which is limited by codes to force and energy maximums, for example, 1.5 seconds opening time and 2.5 seconds closing time); *door fully opened (dwell) time* measured with a stopwatch from inside the elevator as it travels in response to corridor calls (time will vary between 3 and 7.5 seconds, the setting dependent on the distance from the elevator to the corridor call button, as well as the type of call to which the elevator is responding); *elevator "car start" time* measured from inside the elevator with a stopwatch and defined as the elapsed time between when the door has fully closed

in response to a call and when the elevator motion begins (time should be about $^1/_2$ to 5 seconds on traction elevators but could be 2 or 3 times longer on hydraulic elevators); *elevator movement time* ("brake to brake time") to move one floor measured from inside the elevator using a stopwatch and timing "when the elevator starts moving to when it stops moving" (time will be about 5 seconds for a traction elevator with floor heights of 12 ft, but at least twice as long for hydraulic elevators). All of the above can be combined into overall performance time, which is the time from the starting of the door closing at a floor and ending at the time the door reaches the fully opened position at the next floor down or up.

- *Elevator group performance* (measured periodically by the elevator technician). This indicates how quickly a group or bank of elevators responds to corridor calls and lobby service demands during peak periods as well as off-peak periods. This measurement is done most efficiently using an elevator traffic analyzer. Many types are available on the market today, and some later microprocessor elevator supervisory systems have built-in analyzers. A typical printout is shown in Fig. 8.6. Comparisons from one period should be similar in corridor waiting times when compared to other periods. A rise in long waiting times indicates that one or more of the systems features need to be serviced.

### Callback and downtime performance

This measurement reflects maintenance efficiency as well as the overall condition of the equipment. Facility department personnel must work together developing plans to reduce callbacks. The starting point is to know and communicate the volume and causes of callbacks. Having a callback code to easily identify the most common callback causes will be a necessary step in callback reduction. Codes permit easy entry by the technician and facilitate sorting callback cumulative data. The code should not be so detailed as to discourage the use by technician, dispatcher, or build-

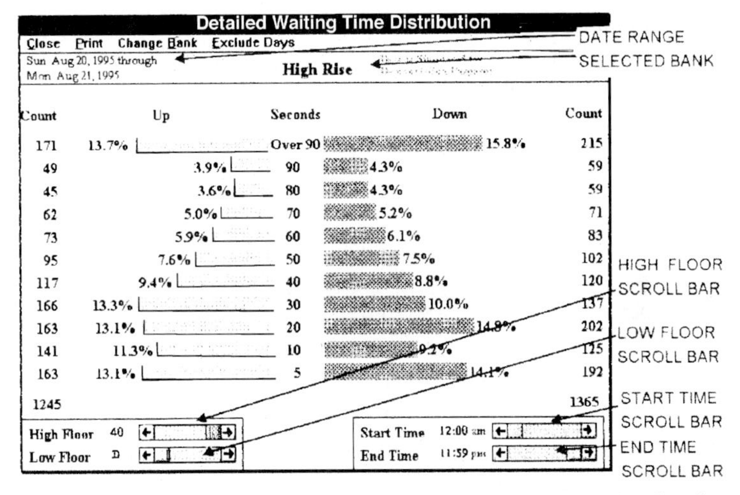


This depiction shows us the overall percentage of answered hall calls. In the illustration above;

- 13.1% of the up hall calls were answered in 5 seconds or less.
- 14.1% of the down hall calls were answered in 5 seconds or less.

**Figure 8.6** Distribution detailed. (*Courtesy of Integrated Display Systems, Inc.*)

ing watch personnel. It should identify an assembly, not a component. It must be written generically. It must red flag unsafe conditions, e.g., entrapments, miss levels, broken steps, stopped handrail. Callbacks of the same type or causes can then be reduced. Intensified maintenance on affected components is one of numerous methods for reducing or eliminating a recurring callback. An effective coding process will help identify equipment components requiring more intensive maintenance or system upgrades. But, like anything, it's only effective if it's accepted and used.

What is the *acceptable* number of callbacks? What is the *achievable* number of callbacks? Answers to the above questions, while highly circumstantial, can be made with some assumptions.

The *acceptable* number of callbacks for a given period should be better than (fewer callbacks) the previous period, or at least the same in cases where the trend has been highly favorable.

An *achievable* number of callbacks for reliably designed equipment could be as follows:

Escalators: two per year (180 days)

Hydraulic: three per year (120 days)

Traction: four per year (90 days)

Another helpful practice is to log the time that callbacks occur, through the workday or at nights or weekends.

Downtime is the number of hours that the equipment is *not* available for passenger or freight usage. A simple chart tracking callbacks and downtime is shown in Fig. 8.7. Observing the trend monthly will track degradation in callback and downtime performances.

#### Monitoring maintenance performance

Your maintenance management plan can provide source data for developing a chart that provides the elevator technician's maintenance performance. Certain tasks are to be performed at given frequencies of time and/or usage. Are they? A chart similar to the one shown in Fig. 8.5 will provide the answer. Figure 8.8 logs maintenance activity, and the Pf column computes the ratio of when the tasks were performed and when they were scheduled. A Pf value of 80 or 90 may be acceptable. On the other hand, a Pf value of 50 implies that maintenance tasks in that category are being performed only one half as often as the MMP promises. This indicates that either steps should be taken to accelerate maintenance activity or the MMP should be revised. Reviewing the chart at least every 3 months is suggested. While this is another piece of paper to generate, the document can be invaluable in defending a "negligence of maintenance" lawsuit. Using a CMMS for monitoring maintenance activity is the most efficient method for documenting maintenance activity. The maintenance technician displays the task or test on a laptop computer or hand-held terminal, then enters the task as completed. The entry documents what maintenance tasks

Area______ Supt______
Route______ Bldg______
Bank______

Callback Activity Log

Report Date __________

| Unit #1 | | | Unit #2 | | | Unit #3 | | | Unit #4 | | | Unit #5 | | | Unit #6 | | | Unit #7 | | | Unit #8 | | |
|---|---|---|---|---|---|---|---|---|---|---|---|---|---|---|---|---|---|---|---|---|---|---|---|
| Date | Time | Act | Date | Time | Act | Date | Time | Act | Date | Time | Act | Date | Time | Act | Date | Time | Act | Date | Time | Act | Date | Time | Act |
| 10/29/97 | 10A | RCF | 10/29/97 | 10A | FUS | 10/29/97 | 10A | TCAB | 10/29/97 | 10A | HEAT | 10/29/97 | 10A | TCAB | 10/29/97 | 10A | HEAT | 10/29/97 | 10A | RCF | 10/29/97 | 10A | FUS |
| 10/29/97 | 4P | DLM | 11/02/97 | 11A | BRU | 10/30/97 | 9A | LH | 11/05/97 | 2P | COMB | 11/05/97 | 2P | LH | 11/05/97 | 2P | COMB | 11/05/97 | 2P | DLM | 11/05/97 | 2P | BRU |
| 11/16/97 | 3P | DLE | 11/16/97 | 10A | O/L | 11/16/97 | 11A | LUB | 11/16/97 | 12N | STEP | 11/16/97 | 1P | LUB | 11/16/97 | 2P | STEP | 11/16/97 | 3P | DLE | 11/16/97 | 4P | O/L |
| 11/24/97 | 10A | DOA | 11/24/97 | 10A | WIRE | 11/24/97 | 10A | HDWE | 11/24/97 | 10A | H/R | 11/24/97 | 10A | HDWE | 11/24/97 | 10A | H/R | 11/24/97 | 10A | DOA | 11/24/97 | 10A | WIRE |
| 12/28/97 | 4P | SBP | 12/28/97 | 4P | CADJ | 12/28/97 | 4P | LEAK | 12/28/97 | 4P | ROA | 12/28/97 | 4P | LEAK | 12/28/97 | 4P | ROA | 12/28/97 | 4P | SBP | 12/28/97 | 4P | CADJ |
| 01/30/98 | 3P | O/S | 01/30/98 | 3P | PCB | 01/30/98 | 3P | NOIS | 01/30/98 | 3P | OTH | 01/30/98 | 3P | NOIS | 01/30/98 | 3P | OTH | 01/30/98 | 3P | O/S | 01/30/98 | 3P | PCB |
| 01/30/98 | 4P | MPF | 01/30/98 | 4P | SAFO | 01/30/98 | 4P | LOW | 01/30/98 | 4P | OTH | 01/30/98 | 4P | LOW | 01/30/98 | 4P | OTH | 01/30/98 | 4P | MPF | 01/30/98 | 4P | SAFO |
| | | | | | | | | | | | | 02/15/98 | 2P | | | | | | | | | | |

**Figure 8.7** Service log. (*Courtesy of Integrated Display Systems, Inc.*)

Area_______ Supt_______
Route______ Bldg_______
Bank_______

Maintenance Activity Log

Report Date __________

| Unit #1 | | | | Unit #2 | | | | Unit #3 | | | | Unit #4 | | | |
|---|---|---|---|---|---|---|---|---|---|---|---|---|---|---|---|
| Date | Time | Act | Pf | Date | Time | Act | Pf | Date | Time | Act | Pf | Date | Time | Act | Pf |
| 10/29/97 | 10A | RE | 82 | 10/29/97 | 10A | RE | 82 | 10/29/97 | 10A | RE | 82 | 10/29/97 | 10A | RE | 82 |
| 10/29/97 | 4P | PP | 62 | 11/02/97 | 11A | PM | 68 | 10/30/97 | 9A | PM | 73 | 11/05/97 | 2P | PM | 80 |
| 11/16/97 | 3P | PM | 79 | 11/16/97 | 10A | PP | 79 | 11/16/97 | 11A | PP | 79 | 11/16/97 | 12N | PP | 79 |
| 11/24/97 | 10A | RE | 82 | 11/24/97 | 10A | RE | 82 | 11/24/97 | 10A | RE | 82 | 11/24/97 | 10A | RE | 82 |
| 12/28/97 | 4P | RE | 69 | 12/28/97 | 4P | RE | 69 | 12/28/97 | 4P | RE | 69 | 12/28/97 | 4P | RE | 69 |
| 01/30/98 | 3P | RE | 67 | 01/30/98 | 3P | RE | 67 | 01/30/98 | 3P | RE | 67 | 01/30/98 | 3P | RE | 67 |
| 01/30/98 | 4P | PP | 62 | 01/30/98 | 4P | PP | 62 | 01/30/98 | 4P | PP | 62 | 01/30/98 | 4P | PP | 62 |
| AVERAGE TOTAL | | | 72 | AVERAGE TOTAL | | | 73 | AVERAGE TOTAL | | | 73 | AVERAGE TOTAL | | | 74 |

| Unit #5 | | | | Unit #6 | | | | Unit #7 | | | | Unit #8 | | | |
|---|---|---|---|---|---|---|---|---|---|---|---|---|---|---|---|
| Date | Time | Act | Pf | Date | Time | Act | Pf | Date | Time | Act | Pf | Date | Time | Act | Pf |
| 10/29/97 | 10A | RE | 82 | 10/29/97 | 10A | RE | 82 | 10/29/97 | 10A | RE | 82 | 10/29/97 | 10A | RE | 82 |
| 11/05/97 | 2P | PM | 67 | 11/05/97 | 2P | PM | 75 | 11/05/97 | 2P | PM | 60 | 11/05/97 | 2P | PM | 75 |
| 11/16/97 | 1P | PP | 79 | 11/16/97 | 2P | PP | 79 | 11/16/97 | 3P | PP | 79 | 11/16/97 | 4P | PP | 79 |
| 11/24/97 | 10A | RE | 82 | 11/24/97 | 10A | RE | 82 | 11/24/97 | 10A | RE | 82 | 11/24/97 | 10A | RE | 82 |
| 12/28/97 | 4P | RE | 69 | 12/28/97 | 4P | RE | 69 | 12/28/97 | 4P | RE | 69 | 12/28/97 | 4P | RE | 69 |
| 01/30/98 | 3P | RE | 67 | 01/30/98 | 3P | RE | 67 | 01/30/98 | 3P | RE | 67 | 01/30/98 | 3P | RE | 67 |
| 01/30/98 | 4P | PP | 62 | 01/30/98 | 4P | PP | 62 | 01/30/98 | 4P | PP | 62 | 01/30/98 | 4P | PP | 62 |
| 02/15/98 | 2P | PM | 67 | | | | | | | | | | | | |
| AVERAGE TOTAL | | | 72 | AVERAGE TOTAL | | | 74 | AVERAGE TOTAL | | | 72 | AVERAGE TOTAL | | | 74 |

**Figure 8.8** The 10 longest wait times. (*Courtesy of Integrated Display Systems, Inc.*)

were completed, when, and by whom. The program compares the date due versus the date completed, permitting a performance log. A microprocessor stores the information, perhaps for years when battery backups are provided, and when equipped with a modem can transmit the information periodically to a central station.

Remote monitoring systems are readily available in today's market and are the state-of-the-art peripheral for today's vertical transportation. These systems can be adapted to any type of equipment. In addition to providing maintenance documentation and performance information as described above, remote monitors can transmit live data on the current state of the elevator or escalator system, similar to the information on an automobile's dashboard but with much more detail. Repeated faulty operations, equipment status, alarms, entrapments as well as system performance information can be transmitted, as they occur, to the central watch station computer. Some monitoring systems have two-way communication to the elevator, helpful for consoling entrapped passengers and deterring equipment misuse. Shown in Fig. 8.9 is a typical remote monitoring system, employing monitors in the security officer's room, at the desk of the lobby attendant, and in the machine rooms for use by the technician. Figures 8.9 through 8.16 provide valuable information to all parties and can be programmed to alert them when certain performance parameters are exceeded. Figures 8.17 through 8.25 illustrate several methods for displaying corridor call waiting times. This information documents how efficiently the elevator dispatching or supervisory system is performing. Figure 8.8 provides status information so that a comparison can be made with the time of day the elevators may be performing less efficiently and the number of cars in service at that time.

#### Improving safety and limiting liability

This section is devoted to reducing possible causes of accidents to building personnel as well as the riding public. No transportation equipment and its associated machinery will be free

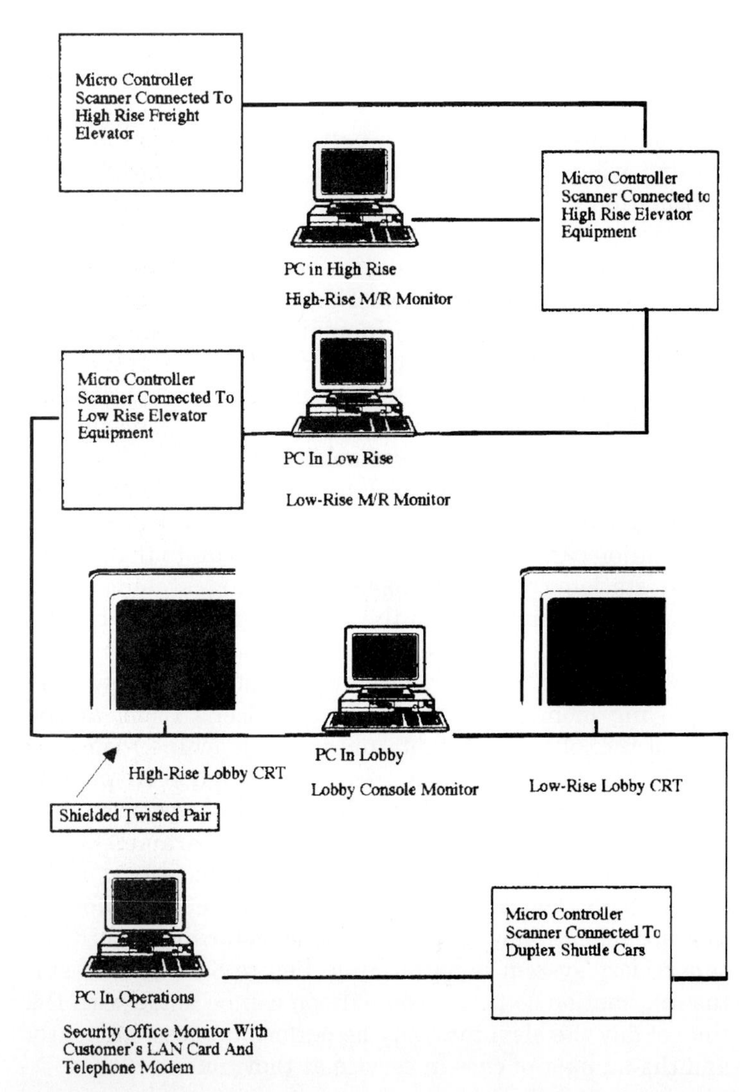


**Figure 8.9** Remote monitoring system. (*Courtesy of Integrated Display Systems, Inc.*)

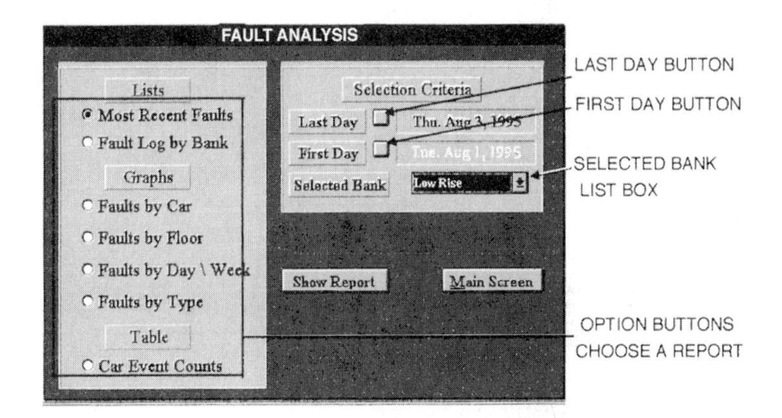


**Lists**
*Most Recent Faults* a list of the most recent faults.
*Fault Log by Bank* a list of faults within a date range.

**Graphs**
*Faults by Car* a graph showing the number and type of faults per car.
*Faults by Floor* a graph showing the number of faults per floor.
*Faults by Day\Week* a graph with the number of faults by days or weeks.
*Faults by Type* a graph with the number of faults per type.

**Table**
*Car Event Counts* a table of minor elevator system failures.

**Selection Criteria ( see pg. 4.9)**
*Last Day Button* choose the last day for the report.
*First Day Button* choose the starting day for the report.
*Selected Bank List Box* select a bank to report.

*Show Report* to bring the selected report on screen.
*Main Screen* to return to the main screen.

**Figure 8.10** Fault report menu. (*Courtesy of Integrated Display Systems, Inc.*)

of potential dangers, no matter how well maintained, no matter how extensive the training programs and cautionary signage. No matter how compliant with safety codes, accidents will occur. How well you have done your job with respect to the issues addressed in the first sections of this chapter will help improve safety in buildings. But accidents will occur, the following being among the most common or serious:

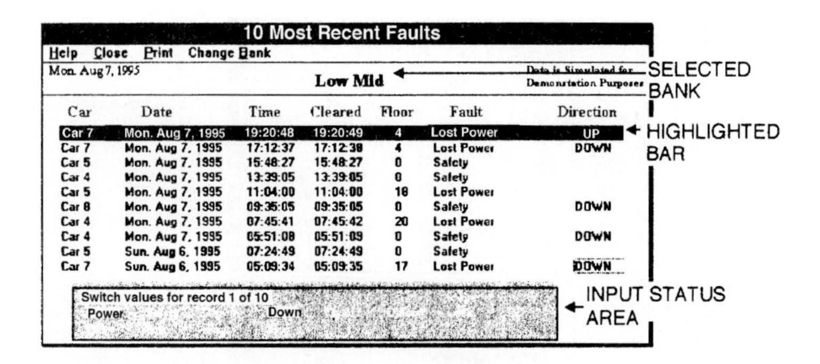


| Car | Date | Time | Cleared | Floor | Fault | Direction |
|---|---|---|---|---|---|---|
| Car 7 | Mon. Aug 7, 1995 | 19:20:48 | 19:20:49 | 4 | Lost Power | UP |
| Car 7 | Mon. Aug 7, 1995 | 17:12:37 | 17:12:38 | 4 | Lost Power | DOWN |
| Car 5 | Mon. Aug 7, 1995 | 15:48:27 | 15:48:27 | 0 | Safety | |
| Car 4 | Mon. Aug 7, 1995 | 13:39:05 | 13:39:05 | 0 | Safety | |
| Car 5 | Mon. Aug 7, 1995 | 11:04:00 | 11:04:00 | 18 | Lost Power | |
| Car 8 | Mon. Aug 7, 1995 | 09:35:05 | 09:35:05 | 0 | Safety | DOWN |
| Car 4 | Mon. Aug 7, 1995 | 07:45:41 | 07:45:42 | 20 | Lost Power | |
| Car 4 | Mon. Aug 7, 1995 | 05:51:08 | 05:51:09 | 0 | Safety | DOWN |
| Car 5 | Sun. Aug 6, 1995 | 07:24:49 | 07:24:49 | 0 | Safety | |
| Car 7 | Sun. Aug 6, 1995 | 05:09:34 | 05:09:35 | 17 | Lost Power | DOWN |

At the top of the scrolling fault window are the following headings:

- ***Car*** which identifies the car at fault
- ***Date*** and ***Time*** the fault occurred
- ***Cleared*** the time the fault was corrected
- ***Floor*** what floor the car was on when the fault occurred
- ***Fault*** the type of fault
- ***Direction*** the direction the car was traveling.

**Figure 8.11** Most recent faults. (*Courtesy of Integrated Display Systems, Inc.*)

- Trips and falls while entering and exiting the elevator or while riding an escalator
- Attempting to escape from an elevator that is stopped between floors
- Being struck by elevator doors
- Being jarred by an elevator making a sudden stop
- Entrapment on the sides, backs, or front of the escalator steps
- Falling into escalator steps
- Falling down an open hoistway

The following sections suggest possible causes for these types of accidents.

1. Elevator accidents involving tripping and falling while entering or exiting the elevator cab can occur as the elevator approaches the floor or when the elevator stops above or

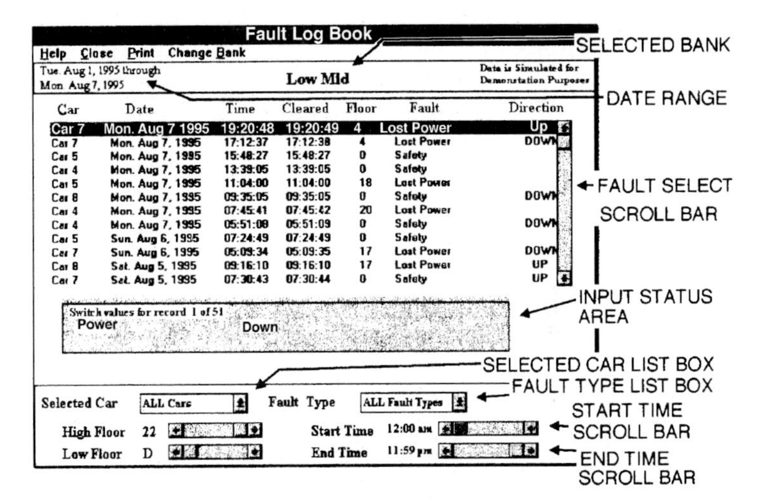


At the top of the scrolling fault window are the following headings:

- ***Car*** which identifies the car at fault
- ***Date*** and ***Time*** the fault occurred
- ***Cleared*** the time the fault was corrected
- ***Floor*** what floor the car was on when the fault occurred
- ***Fault*** the type of fault
- ***Direction*** the direction the car was traveling.

**Figure 8.12** Fault log by bank. (*Courtesy of Integrated Display Systems, Inc.*)

below the corridor floor. Many traction elevators open their doors prior to reaching floor level. This is referred to as "pre-opening the doors on approach" and is done to improve performance times. Where preopening is provided, the doors should be less than 24″ open when the elevator is within $^1/_2$″ from floor level, i.e., less than the width of a person (24″ wide). Observe the approaches as you use the elevator and report any faulty operations to your elevator technician. Tripping and falling can also occur when the elevator stops $^1/_2$″ beyond floor levels. Point out any of these faulty operations to your elevator technician.

2. Elevators stopping above or below floor level more than 2 ft will not normally open its doors. If they are forced open by an entrapped passenger attempting to exit

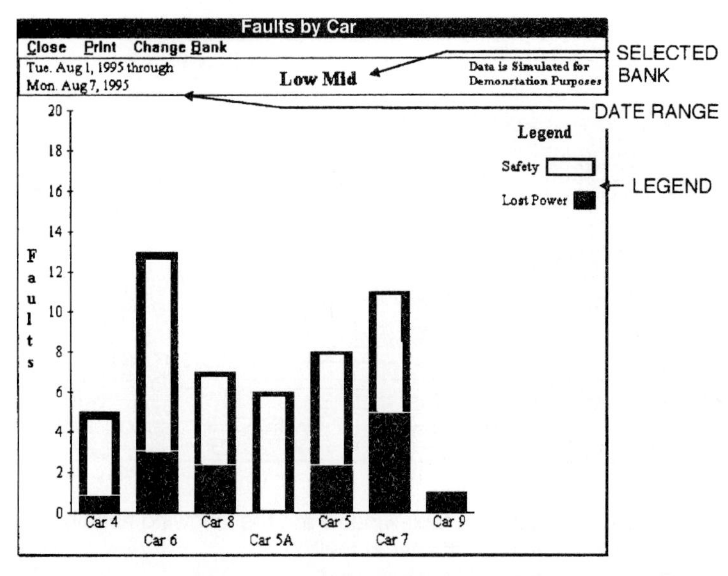


**Figure 8.13** Fault graph by car. (*Courtesy of Integrated Display Systems, Inc.*)

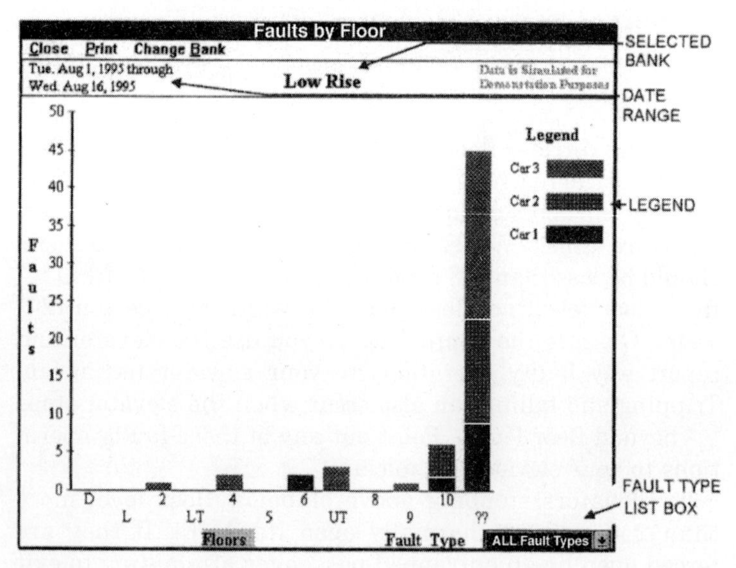


**Figure 8.14** Fault graph by floor. (*Courtesy of Integrated Display Systems, Inc.*)

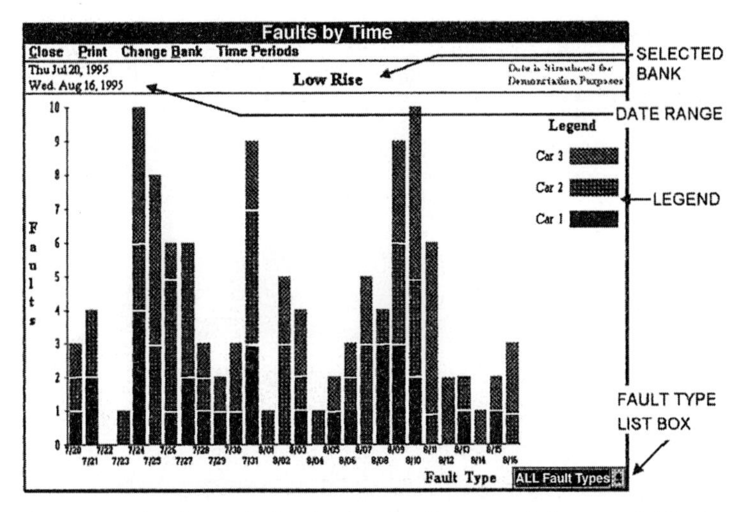


**Figure 8.15** Fault graph by day/week. (*Courtesy of Integrated Display Systems, Inc.*)

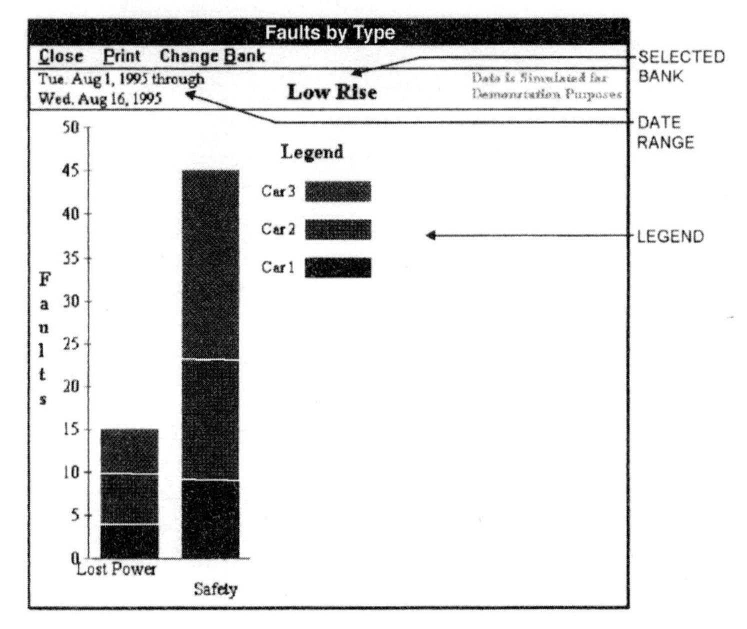


**Figure 8.16** Fault graph by type. (*Courtesy of Integrated Display Systems, Inc.*)

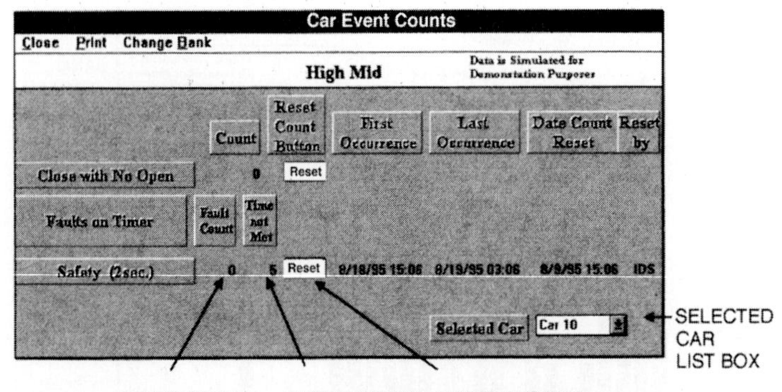


In the example above, Car 10 has had 5 safety circuit faults which were cleared within 2 seconds between 8/18/95 and 8/19/95.

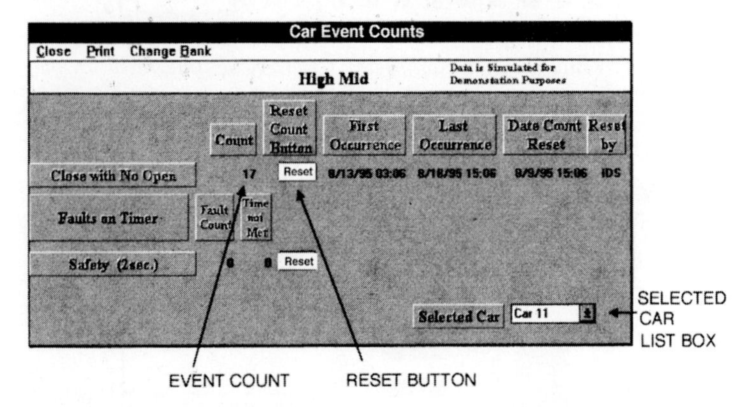


In the example above Car 11 has had 17 events in which the door close limit was active without a corresponding door open limit. This could be due to an intermittent contact or other door malfunction.

**Figure 8.17** Car event counts. (*Courtesy of Integrated Display Systems, Inc.*)

the elevator, accidents can occur. Falls down the hoistway sometimes occur when passengers jump from the elevator that has stopped 3 ft or more above the floor, lose their balance, and fall into the hoistway. Provide warnings to your tenants. Explain that an entrapped passenger must wait for an authorized person to release them from the elevator, and explain how they contact that person using

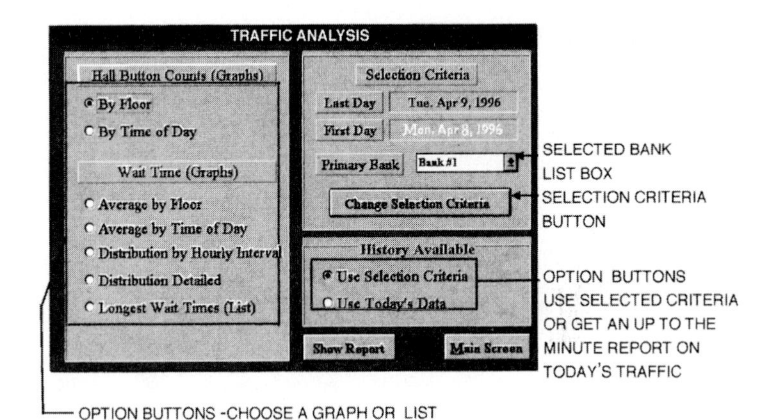


**Hall Button Counts (Graphs)**

*By Floor* a breakdown of hall button density per floor.

*By Time of Day* for a 24 hour timeline of hall button density at 10, 15, 20, 30, or 60 minute intervals.

**Wait Times (Graphs)**

*Average By Floor* to see the average waiting time per floor.

*Average By Time of Day* for a 24 hour timeline of average waiting times at 10, 15, 20, 30, or 60 minute intervals.

*Distribution By Hourly Interval* to get 24 hour timeline showing the percentage of calls answered in 5, 10, 20 ... 90 seconds or longer.

*Distribution Detailed* for a more detailed view of the percentages.

*Longest Wait Times (List)* to produce a list of the floors with the 10 longest waiting times.

**Figure 8.18** Traffic report menu. (*Courtesy of Integrated Display Systems, Inc.*)

the alarm and communication system. The passengers should attempt to move the elevator only with the floor buttons and should stay clear of the doors.

3. Most elevator passengers are aware that the doors close automatically. Most elevator passengers assume that the door reopening devices, e.g., safety edges and light rays, are working properly. However, these devices are not usually effective on the corridor doors, unbeknownst to the average passenger. Instruct them of this fact. Also do not circumvent any door reopening device. If a device is not working properly, alert your technician. Also have your

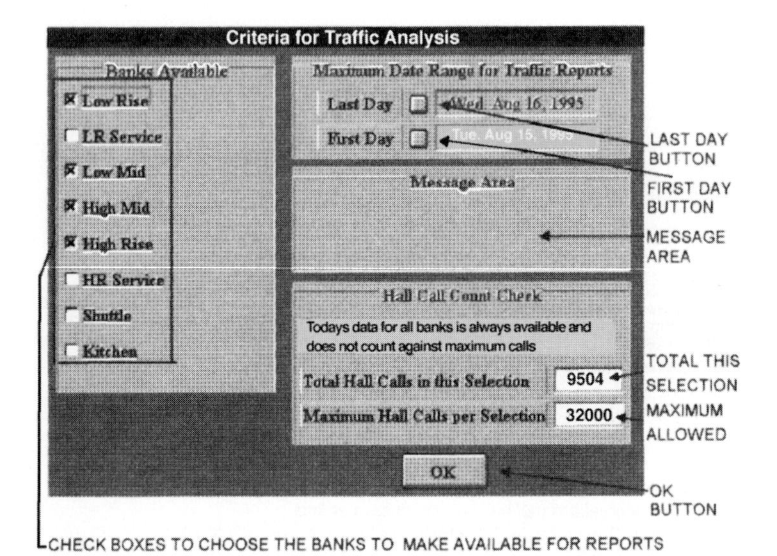


**Figure 8.19** Selection criteria menu. (*Courtesy of Integrated Display Systems, Inc.*)

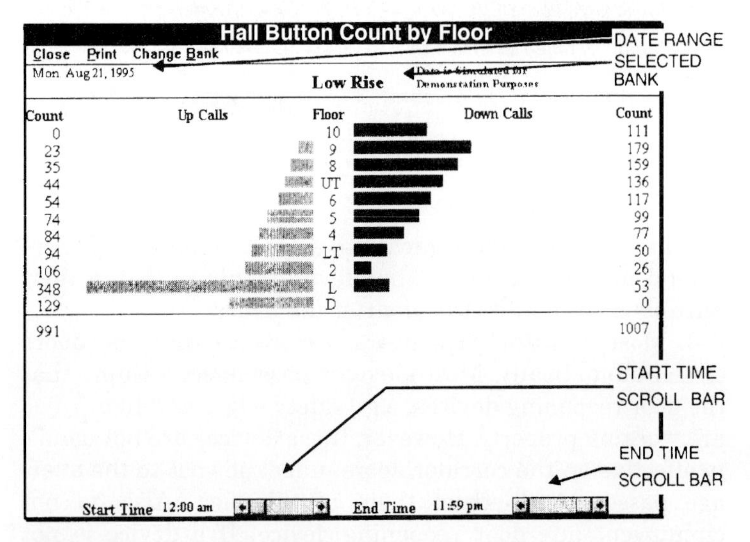


**Figure 8.20** Hall button count by floor. (*Courtesy of Integrated Display Systems, Inc.*)

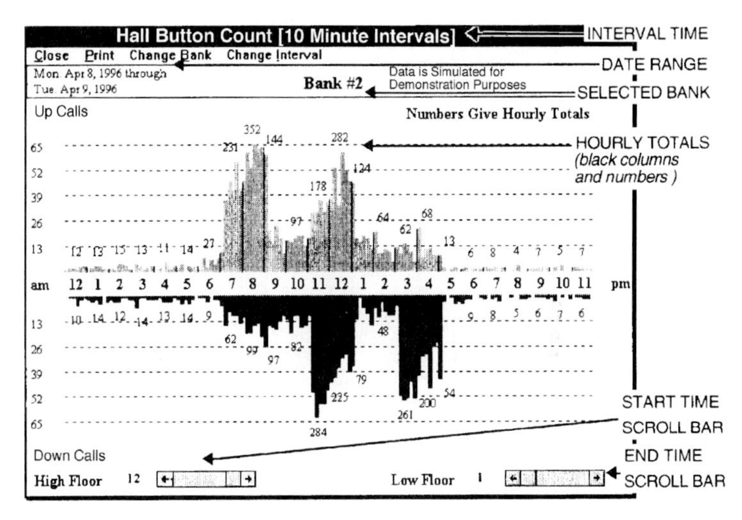


**Figure 8.21** Hall button count by time of day. (*Courtesy of Integrated Display Systems, Inc.*)

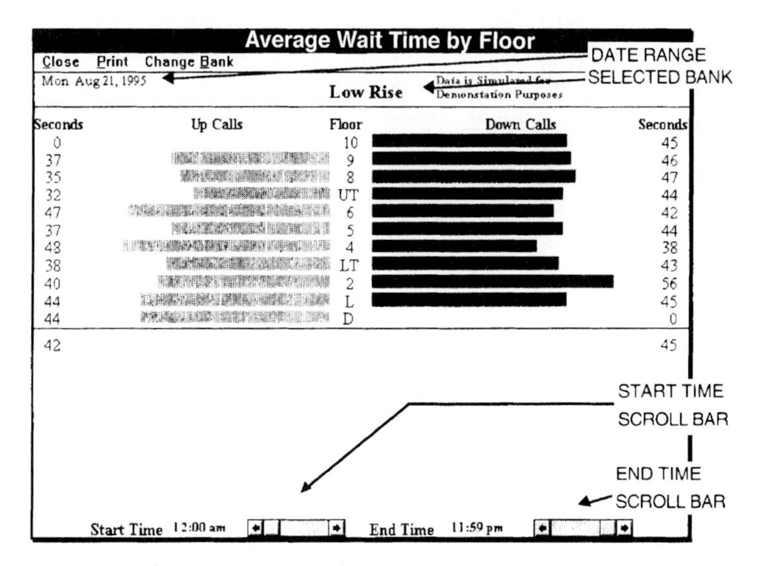


**Figure 8.22** Average wait time by floor. (*Courtesy of Integrated Display Systems, Inc.*)

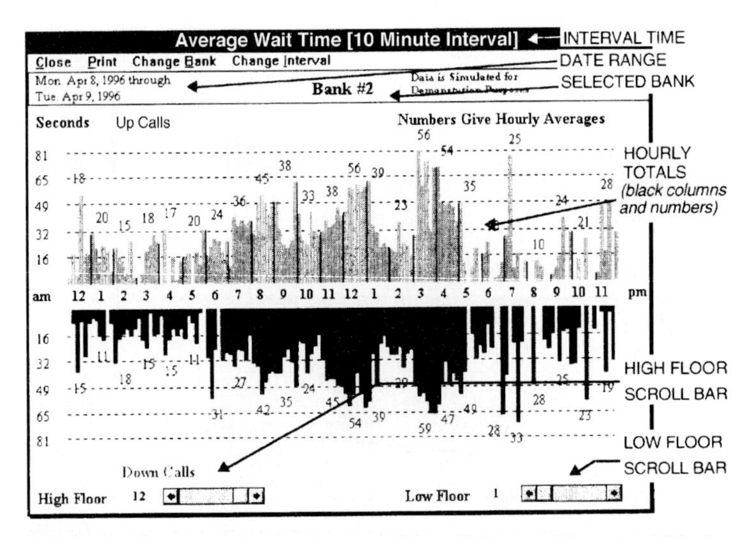


**Figure 8.23** Average wait time by time of day. (*Courtesy of Integrated Display Systems, Inc.*)

technician demonstrate how to test the doors to assure that they comply with both the force and energy codes. The force code is written to reduce the squeezing force of the doors to a tolerable level (80 foot pounds) should the closing doors trap someone. This is measured with a force gauge. The energy code (not to exceed 7 foot pounds) is written to limit the striking force of the door system. Testing is done with a stopwatch. Elapsed door closing time should not be less than a value that would indicate the door speed has reached a maximum (usually 1 ft/s for standard door system weights). For example, a 3-ft-wide single panel door should take at least 3 to 5 seconds to close, perhaps longer for heavier doors such as ornamental bronze doors.

4. Falls down elevator hoistways have occurred. They can occur because of circumventing of the electrical and mechanical interlock on the hoistway door, unauthorized personnel using a door interlock releasing tool, broken relating mechanisms on multipaneled hoistway doors, failure of a mechanical door closer, unsecured top or bottom door retaining devices, or simply when excessive force is applied

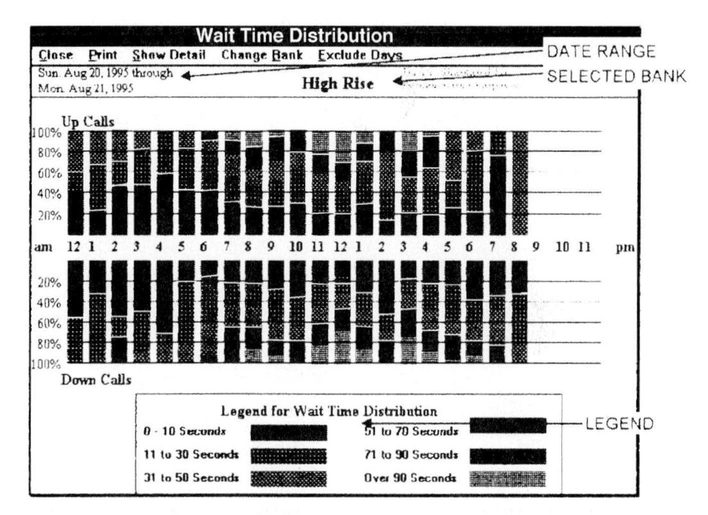


**This graph shows a 24 hour time line with the percentage of wait time for every hour.**

For example, between the hours of 8pm and 9pm;

- 100% of the up calls were answered in 31 to 50 seconds.

The down calls for that same time period read as follows;

- 36% were answered in 0 to 10 seconds.
- 44% were answered in 11 to 30 seconds.
- 20% were answered in 31 to 50 seconds.

See Distribution Detailed *(pg. 5.7)* for a breakdown of this graph's percentages.

If waiting times are unusually long, check the *FAULT LOG* and the *SERVICE LOG* to see if 1 or more cars were out of service for maintenance or malfunction at the time the waiting was the longest.

**Figure 8.24** Wait time distribution. (*Courtesy of Integrated Display Systems, Inc.*)

on the corridor side of the hoistway door. Thus reduce the risk of this happening in your buildings, frequent and thorough hoistway door maintenance is necessary. Door interlocks, hangers, sills, bottom shoes (gibs), relating cables, and closers must be in good condition. Door interlocks should not be circumvented unless absolutely necessary, and then only by an elevator technician when the elevator is *out* of normal service. In the event that a door is opened and the elevator is at another floor, place a barricade in

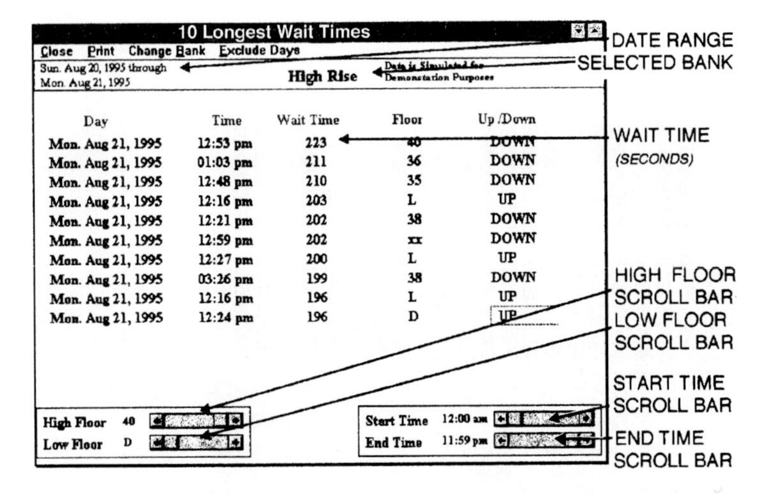


This report shows the ten longest waiting times of the hall buttons pressed. The date, time, floor, and direction are recorded. This can be an accurate and unbiased answer to the complaint, "I WAITED FOREVER FOR AN ELEVATOR ON THE 29TH FLOOR THIS MORNING".

If waiting times are unusually long, check the *FAULT LOG* and the *SERVICE LOG* to see if 1 or more cars were out of service for maintenance or malfunction at the time the waiting was the longest.

**Figure 8.25** Long wait times. (*Courtesy of Integrated Display Systems, Inc.*)

front of the opening and take the elevator out of service until repaired.

5. Hydraulic elevators having leaks or unaccounted oil losses are potentially dangerous. A pipe or cylinder rupture can result in an elevator descending at a rapid speed and contacting the bottom springs with great force. Injury or death is likely. Complete records of any adding of hydraulic oil must be maintained.

6. Falls on escalators can be caused by slower-moving or stopped handrails, sudden stops, transporting of large packages, wheelchairs, carts, and freight, and simply the loss of balance. Instructing passengers to hold the handrails, to avoid the sides of the step, and to step up over the combfingers when exiting the escalator assists in averting loss of balance accidents. Signs should also be placed near an escalator directing wheelchaired passengers and less agile people to

the location of the elevator. Instruct the building tenants accordingly. Shut down any escalator having either or both handrails stopped or moving more slowly than the steps. Flaps or guards should be secured over the stop buttons at each end of the escalator to prevent accidental engaging of the stop button. Escalators are permitted by code to travel at speeds up to 125 ft/min. If you feel this speed is too fast for your building's tenants, consider a slower speed. Some escalators by design have a slow and fast speed selection on the starting key switch. Contact the escalator manufacturer to obtain escalator speed-reducing information. They also have a vested interest in reducing escalator accidents.

7. Entrapments on escalators occurring around the edges of the step can be reduced if passengers stand in the middle of the step. Yellow markings forming a box around the step help to direct passengers to stand in the middle. Consider having these markings applied to your steps. The entrapments are generally caused by excessive gaps between adjacent steps, between steps and the side skirts, and the failure of the comb fingers to mesh into the step treads. Excessive friction on the side skirts increases the chance for entrapments. All of the above should be minimized. Side gaps should be limited to $^3/_4''$ or less on both sides of the step throughout the escalator travel but especially near the ends, where the steps go from offset to level. If scraping noises are heard while trying to comply with these clearances, applying a plastic-like material to the sides of the steps can also reduce the gap. Contact your elevator parts supplier or your consultant for more information on this product. The side skirts should have less rubbing friction than the surrounding metal surfaces to help reduce entrapments. If not, apply a friction-reducing substance either permanently or periodically to the side skirts, being careful not to get any of the substance on the top of the steps or on the landings. Serious injuries or deaths can occur when step frames and/or rollers are cracked or broken. Annual internal inspections of the steps are a must to help ensure the structural integrity of the steps. Figure 8.26 illustrates typical escalator components.

8. Exposing unauthorized personnel to vertical transportation machinery rooms can be hazardous. Keep the rooms locked and provide entry to only authorized personnel. Do not use these rooms for storage of equipment other than maintenance equipment and parts needed for vertical transportation. This includes the elevator hoistways and pits. These areas contain dangerously high voltages and many moving parts. Cleaning personnel, painters, and other people who have no operational equipment in these areas should not be allowed to use the areas. In addition to personal injury and death hazards, equipment damage can occur. For example, the storing of large objects in the pit underneath the counterweight area can cause frame damage resulting in the dislodging of the counterweight from their containment rails. If as a result a collision were to occur between the car and counterweight, it could destroy the elevator cab and injure or kill its occupants. This is not a doomsayer's fantasy; these collisions have occurred.

Elevator technician safety can be enhanced with proper lighting in all working areas. Included are the machine rooms, the car tops, and the pits. Yellow striping on low overheads, ledges, and at pinch points will assist the technician in being aware of these dangers. High-voltage signage should also be placed where needed. Hoistway access switches will assist the technician to gain safe access to the hoistway. This feature should be added if it does not already exist. Ask the technician to suggest other safety measures or point out any potentially hazardous conditions.

Failures to warn of any potentially dangerous conditions have resulted in numerous settlements in favor of the plaintiff. If a code-enforcing authority, elevator technician, or consultant informs you of potential dangers, you are obligated to correct the condition or forewarn the users of the equipment. In like manner if you notice a potentially dangerous condition that should have been detected and corrected by a contracted elevator maintenance company, notify them immediately by registered letter or courier of the condition. If in doubt shut down the equipment until the correction is made.

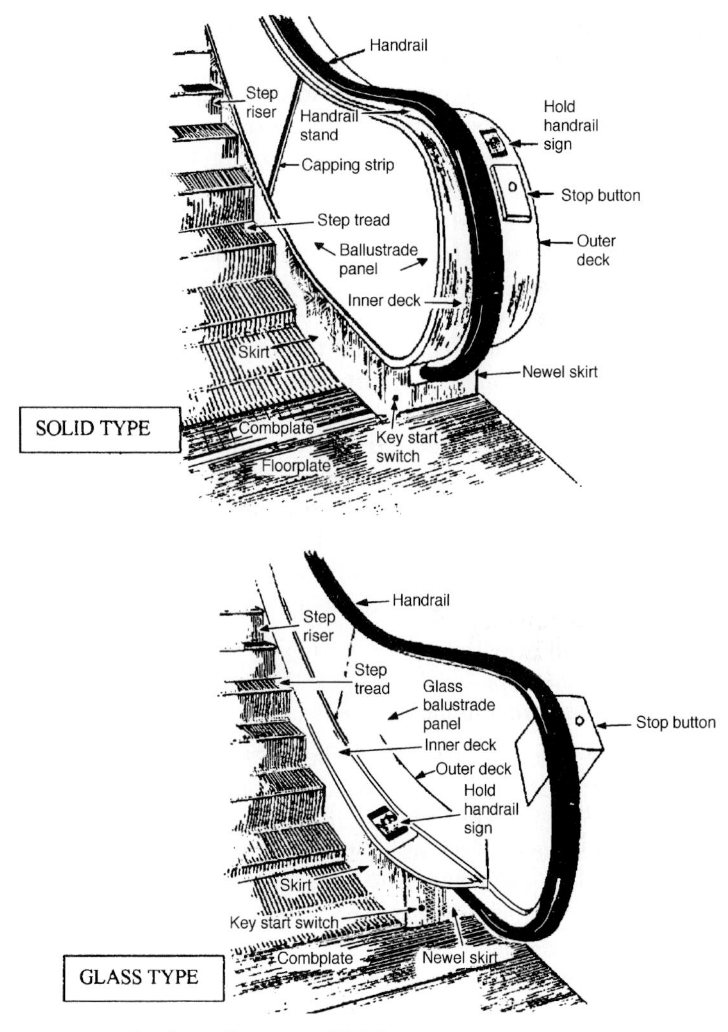


**Figure 8.26** Escalator. (*Courtesy of NEII.*)

### Modernization of equipment

The first significant changes to vertical transportation equipment in the last 50 years occurred in the 1980s. Microprocessor technology was introduced permitting programmable features on both the driving and supervisory systems. Prior to this era, features and options required to optimize equipment performance when a building traffic pattern changed were costly to add and required extensive downtime. System performance information and error messages so valuable to the technician for troubleshooting equipment malfunctions were virtually nonexistent. Reliability was poor. Maintenance was more labor intensive. Safety features on both elevators and escalators were not as abundant as they are today. Noise levels were higher. Ride quality was poor in comparison. Signal lights and gongs operated poorly when compared to modern fixtures. That's the *bad* news if your equipment was built in the 1970s or earlier. The *good* news is that the bulk of the elevator equipment can be reused, and the remaining equipment can be replaced (modernized, updated) in stages. Unfortunately the good news does not apply to escalators. The only good news about escalators is that they can be removed and replaced in less time, and installation labor is not as intensive, provided there is a passageway into the building for a pre-assembled escalator.

### Why, when, and what to modernize

Most of the reasons were listed above. "Keeping up with the Jones's" is another reason. Making your equipment more pleasing to the eye than the equipment in the office building across the street could attract more tenants. But if your equipment is over 20 years of age and has been subjected to heavy usage and/or abused, you may have more than one reason to modernize. Equipment manufactured in the 25 years from 1950 to 1975 contained hundreds of relays and similar devices. The life expectancy is 10 million operations; however, reliability studies usually indicate a replacement at 2 to 3 million operations. Unfortunately most of these devices are hardwired and not readily replaceable, when compared to plug-in devices. Many failures can occur, making the new microprocessor control very attractive. Elevator controllers and energy-efficient drive units may be included in the first stage of modernization. Following close behind are cabs and car stations inside the cab, mostly for aesthetic reasons since the old cab architecture and pushbuttons may look aged and worn. Usually one elevator is modernized at a time, thereby minimizing disruptions in traffic flow. Doors, their operator, and the door safety equipment are next. Door panels can be optional. The cab superstructure, guardrails, counterweight, overhead driving machine, and sheaves usually need not be replaced and can last several decades. Most likely the motor generator (old power source for the drive motor) will not be needed. A solid-state drive will replace it. So, as was stated above, the bulk of the elevator equipment will remain and the updated additions will make the riding public feel that they are using the latest, most advanced elevator system money can buy.

#### How to select a modernization contractor

So, you've decided to modernize some of your equipment in stages, depending on the available budget each year, for the next several years. What do you buy first and what do you need most? We hate to keep harping on this, but contact your elevator consultant and have the firm survey your equipment. If you don't have a consultant, look in the yellow pages or the consultant directory in the *Elevator World Magazine* Source Directory (under the consultant tab). Have the consultant write the specifications for the modernization, assist you with contractor selection, and oversee the installation. All that! Now you know why we mentioned the consultant.

Consultant responsibilities for equipment modernization include:

- Evaluate the equipment in the building. The evaluation should include the equipment condition, current and projected reliability, ride quality, noise performance, and passenger-handling capability. All of the above should be compared with industry standards to highlight any deficiencies. Tests should include motor insulation breakdown, underground cylinder and piping pressure tests for leakage, and a complete structural examination of the escalator's internal parts including all step frames. The final report should indicate test results.

- Determine how many elevators can be modernized at a time. Based on the passenger-handling capabilities above, establish the minimum traffic-handling performance that would be tolerable.
- Establish the modernization stages, how much time each stage will take, which stage should be done first and why, and the projected costs for each stage.
- Write the specifications that will be submitted to a list of prequalified elevator contractors for competitive bidding. The specifications must include equipment changes necessary to meet current local code requirements for the alterations. They must also include performance assurances and the test methods that will be used to assure compliance. Usually there will be patch-up and other work associated with the modernization. The specifications must define what work is the responsibility of the elevator contractor and what other contractors will do. Interim maintenance required on the installed equipment must be included, along with a maintenance contract for the new and existing equipment.
- Rationalize any exceptions to the specifications and determine the impact of the changes.
- Oversee the installation of the new equipment by making periodic job inspections.
- Ensure that specifications have been met for each stage of modernization. If not, make a list of deficiencies (punch list) and a time frame for correcting the deficiencies; then check and verify conformance.
- Assist facility management in determining when final payments should be made for each stage of modernization.

The facility manager's responsibilities prior to and during the modernization include the following actions:

- Notify all tenants of the purpose and starting date.
- Explain that longer wait times for elevator response will occur and ask for their indulgence.
- Ask them to follow good elevator practices, e.g., walk up one floor and down two, do not detain the elevator door from closing, and stagger their working hours if possible.
- If the modernization elevator contractor is not the current maintenance contractor, consider selecting only the modernization contractor to perform maintenance on all equipment during all stages of modernization. Otherwise, personnel issues may occur.

# NOTES

Chapter

# 9

# Water Treatment Services

## Introduction

Facilities depend on water systems for space heating and cooling, for heat removal and environmental control in data base facilities, for humidity control in sensitive environments, and for domestic (potable) water service. These systems range from simple to very large and complex. The equipment may include many different combinations of cooling towers, chiller/evaporators, heat exchangers, direct expansion (DX) chillers, fan coil units, etc. The piping systems include large diameter main risers, smaller diameter lateral distribution piping, and lines carrying water to and from individual heating and cooling units.

Water is a reactive substance. In passing through piping and equipment, it can cause corrosion of metals and/or it can lay down mineral scale deposits. Corrosion can lead to metal failures and leaks. Corrosion products and mineral scales form deposits that can reduce water flow in pipes and block condenser tubes, fan coils and other heat exchange equipment. Microbiological fouling from bacteria and organic matter in the water can also lead to corrosion and loss of heat transfer and water flow capacity.

Water treatment chemicals and services are used to help control these problems and maintain trouble-free operation of the facility heating and cooling systems. Water treatment may include various combinations of chemicals and services provided by a vendor, plus work done by facilities personnel. This work includes applying chemicals, controlling operating systems, chemical testing and reporting of control and performance data.

## Facility Water Systems

This section describes common facility water systems and the problems that can be encountered when operating these systems. Only water-side operations are discussed here. Air-handling equipment and the mechanical aspects of chiller and boiler operations are described elsewhere in this Handbook.

## Air-Conditioning Systems

Figure 9.1 is a schematic diagram of a typical simple building air-conditioning system including a cooling tower, a chiller unit, air-handling units, and condenser water and chilled water circulating systems. Systems such as this are sometimes referred to as heating, ventilating, and air-conditioning (HVAC) systems. Specific systems may be very complex, but in principle all are variations of the equipment and circulating water systems shown in Fig. 9.1. Following are brief descriptions of the component parts of this typical HVAC system.

**Chilled water.** Building air is cooled by circulating the air over small fan coil units in individual rooms, or large air-handling units that deliver cooled air to public spaces and to duct systems. Chilled water circulating in a closed loop (see Fig. 9.1) absorbs heat from the air in these fan coils and air-handling units. The warmed water returns to the chiller machines, where the heat is removed.

Chilled water systems may be small, supplying all or part of single buildings, or they may be very large, servicing

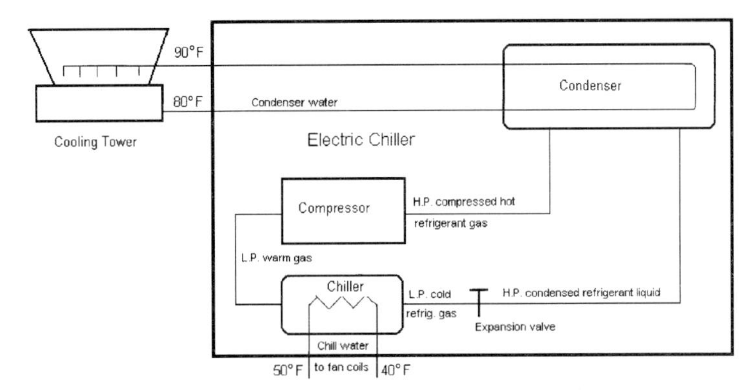


**Figure 9.1** Schematic drawing of a simple building air-conditioning system.

many buildings in a campus facility. Sometimes a single large chilled water loop may be cooled by chillers in several buildings. In campus facilities or in high-rise buildings, secondary chilled water loops may be used to cool specific buildings or tenant spaces. Secondary loops may be separated from the primary chilled water by plate-and-frame heat exchangers, or they may receive direct injection of primary chilled water as needed to maintain the required temperature in the secondary loop.

**Condenser water and cooling towers.** The chiller equipment in an HVAC system contains two shell and tube heat exchangers. The first, called the *chiller,* transfers heat from the circulating chilled water to a refrigerant system. This can be either Freon gas in a compression chiller, or lithium bromide solution in an absorption chiller.

The second heat exchanger in the chiller machine, called the *condenser,* transfers heat from the refrigerant to an open circulating water system called the *condenser water.* In this heat transfer process, the compressed refrigerant gas, if used, is condensed to a liquid.

The cooling tower is the ultimate heat sink in an HVAC system. Heated condenser water falls through the cooling tower and is cooled by evaporation in a stream of air. In natural draft cooling towers, the temperature difference across the tower is sufficient to cause air flow through the tower. Large cooling towers in power stations and industrial plants often use natural draft designs. Most often, induced or forced draft fans push or pull air through the cooling tower.

**Free cooling.** Electric drive compression chillers, commonly used in large facilities, require large amounts of electric power to drive the compressors. In some parts of the country and at some seasons of the year, the condenser water, after passing through the cooling tower, becomes cold enough to be used directly for cooling the building air, with no need for chilled water. Under these conditions, the chillers can be bypassed, and the cold condenser water can be sent directly through the chilled water system to the fan coils and air-handling units.

Bypassing the chillers in this way mixes the condenser water and the chilled water, and leaves condenser water in the chilled water pipes when the system is returned to normal operating mode. This can create severe corrosion, deposition, and microbiological problems in the chilled water system. To avoid these problems, the condenser and chilled water systems are sometimes separated by a plate-and-frame heat exchanger. This eliminates mixing of the system waters, albeit with some loss in system efficiency due to the heat exchanger.

This process is called "free cooling" because the energy cost associated with operating the chillers is eliminated. This is an attractive option, particularly in systems that can operate in free cooling mode for several months each year. However, the energy savings must be balanced against the increased costs of water treatment and the potential for damage to the chilled water system. For this reason, use of a plate-and-frame heat exchanger often becomes an attractive option.

### Heating systems

HVAC heating systems utilize circulating steam or hot water. Steam may be supplied by a local utility station or generated on-site in low-pressure boilers. Large campus facilities and some high-rise buildings make their own steam and drive turbines to generate electric power. Steam extracted from the turbines is used for heating.

In some cases, hot water is circulated for heating purposes. Hot water boilers are similar in appearance and function to firetube steaming boilers, except that no boiling occurs and no steam is generated. Closed hot water systems are similar to closed chilled water loops, and they are treated chemically in the same way, except at higher dosages.

Many different piping configurations are used for heating purposes. In some cases, the same pipes and fan coils are used for chilled water and either hot water or steam. In other systems, piping for the cooling and heating systems may be different, but the same fan coils and air-handling units may be used. When planning a water treatment program, it is important to understand the operation of each system, including possibilities for mixing waters and for introducing air or losing water from otherwise "tight" systems.

### Humidification systems

Hospitals, computer facilities, and other sensitive locations require humidity control in defined parts of the facility. Humidity is controlled by direct injection of steam to the air ducts as required. These systems are controlled automatically by humidity sensors.

In some cases, facility boiler steam is used directly for humidification. This is the simplest and least expensive route to humidity control. However, this method severely limits both the type and amount of additives that can be used to protect the steam and condensate system from corrosion. A better method is to use boiler steam through a heat exchanger to boil potable water or de-ionized water to make clean humidification steam. Simple units that can be installed as part of the air duct system are available for this purpose. If potable water is used for humidification steam, provision must be made to remove the solids deposited from the evaporated water.

## Potable water systems

Facilities may use a municipal water supply, or they may develop a private water supply, usually well water or river water. This supply normally provides both potable (domestic) water and nonpotable (cooling, boiler, and process) water to the facility.

**Municipal water supplies.** In most cases, a municipal water plant can supply properly treated water for all of a facility's needs. This is the simplest, safest, and most reliable source of plant water. However, use of a municipal water supply does not guarantee a trouble-free system. A soft (low hardness) corrosive water can corrode steel pipes and equipment, creating corrosion product (chip scale) problems and red water. Pipe failures can occur, first at threaded joints, welds and other stressed areas, and then as pitting corrosion failures in the system. Soft municipal water supplies can also cause pitting corrosion of copper pipes. At the other extreme, hard water supplies can lay down calcium carbonate deposits in pipes and heat exchangers, that can seriously impede water flow and heat transfer.

**Private water supplies.** A municipal water supply may not always be the most cost-effective source of facility water, although the apparent savings available from using a private supply must be offset by the costs of the required treatment and maintenance responsibilities.

Well water, if available in sufficient quantity and quality, can often be used for potable water simply with chlorination. In other cases, treatment to correct red water or black water problems, and to control calcium carbonate scale, may be required. Some very large facilities may find it cost-effective to use an available surface water as a potable supply. In such cases, additional processes such as clarification, filtration, softening, and color removal may be required. These processes require space, equipment, and personnel for continuous operation, and they will create solid and liquid wastes for disposal. Such plants may require state environmental discharge permits. Any facility considering development of a private water supply should contract with consultants and water treatment vendors to design the treatment plant and estimate required operating personnel and costs.

### Principles of water treatment

This section introduces the basic principles of facility water treatment. More information can be found in excellent reference textbooks prepared by the major water treatment vendor companies, and in technical papers published by vendors, consultants, and users of water treatment.

Water that evaporates in a cooling tower must be replaced by makeup water. Since the dissolved and suspended solids in the water do not evaporate, the circulating condenser water becomes several times more concentrated than the makeup water. The concentrated water may become more corrosive and/or more scale-forming than the makeup, depending on the water composition. Also, air drawn through the cooling tower brings with it soil, construction debris, and various kinds of organic and microbiological matter. In industrial facilities, oil and process contaminants can enter the cooling water. All of these materials can coat pipes and heat transfer surfaces, and collect in the basin. Deposits formed in this way become breeding grounds for organisms that can cause slime formation and under-deposit corrosion, called *microbiologically influenced corrosion* (MIC).

Closed chilled water and hot water systems should remain clean, and if properly treated, free of corrosion and mineral scale. However, any water lost from these systems must be replaced, and the makeup water brings with it opportunities for contamination as described above. Expansion tanks on closed systems are often open to the atmosphere, providing an entrance point for oxygen, airborne debris, and bacteria. Hot water systems are especially prone to oxygen pitting corrosion. Under-deposit localized pitting, sometimes involving bacteria, is the most common form of corrosion in closed systems.

All of these problems are controlled by a combination of water treatment and good operating practices. There are three cardinal principles for successful water treatment in any facility. These are: circulate the water, provide proper levels of appropriate water treatment, and control the system within established operating parameters. Each of these principles is discussed briefly below.

**Circulate the water.** With few exceptions, systems with moving water will be in better condition than similar systems in which the water circulates only occasionally if at all. Moving water provides fresh supplies of treatment chemicals, carries away waste products, and helps to prevent hard deposits from forming in the system.

Unfortunately, not all systems can be circulated continuously. Hot water systems are layed up in the summer, and some condenser water and chilled water systems in the winter. Sometimes these systems can be circulated occasionally without a heat or cooling load, but in other systems this is not possible. Dead legs never see circulation except when portions of the system are drained.

Solutions to circulation problems that prevent proper water treatment must be developed cooperatively by the water treatment vendor and facilities personnel. Possibilities include intermittent regular circulation; high-level chemical treatment with replenishment as needed; dry lay-up; and installation of bypass lines to make circulation possible. All of these and other possibilities must be considered. Systems that are simply closed up and ignored for several months will deteriorate and eventually cause serious and expensive problems.

**Provide proper chemical treatment.** Corrosion and mineral scale formation can never be completely eliminated in operating water systems, except at unreasonable cost. However, these problems can be controlled at levels that do not affect system operations and that do not seriously reduce the availability and life of the equipment.

The key to success is good quality water treatment. The nature and amount of water treatment that is required in any facility is site specific. It depends on the facility, the type of equipment involved, the heating and cooling loads, and the nature and availability of the water supply. Water treatment can vary from simple chlorination of once-through cooling water to complex chemical treatment of high-cycle cooling towers and high-pressure boilers. Some systems can, in fact, operate without chemical treatment, but this is not generally the case. Water treatment decisions should be made on the advice of experienced water treatment vendors and consultants.

**Control the systems.** In addition to providing good chemical treatment programs, it is important to control the dosage levels of chemicals and the operating parameters of the system. These can include, for example, the pH and conductivity of the water, required minimum flow rates, and so forth.

The water treatment vendor will establish minimum and maximum control ranges for each treatment chemical. These ranges are important. Exceeding the maximum limit will unnecessarily increase the cost of chemical treatment and may, in some cases, cause adverse chemical reactions to occur. The minimum of the control range is the lowest level at which the chemical can be expected to perform properly. If a chemical dosage falls repeatedly below the lower limit, protection may be lost. Corrosion that occurs under such circumstances cannot be reversed, and deposits that form may require off-line cleaning with strong chemicals for removal before chemical treatment can again be effective. Also, simply returning a chemical dosage to the control range may not be sufficient to reestablish protection. High-level treatment may be required to establish protective films that can then be maintained by dosages within the control range.

For these reasons, chemical feeding and control equipment should be an integral part of the total chemical treatment program. While complete manual control of water treatment systems is possible, some degree of automation provides improved reliability with fewer demands on personnel. Simple control systems include timed additions of chemicals and bleed based on conductivity. More advanced technology can include chemical feed based on makeup water meter readings, sophisticated on-line analytical data, and precise-dosage pumps. Remote monitoring and control capabilities allow vendors and facility personnel to read data and make system adjustments from off-site locations. Specific feed and monitoring equipment must be selected for each facility, and perhaps for each system within a facility, to provide the necessary degree of control at that site.

### Obtaining water treatment services

This section deals with the types of water treatment services that are available, and with the process of selecting and installing a new water treatment vendor at a facility.

### Contracted services versus in-house water treatment

Water treatment vendors provide both the required chemical products and necessary service to apply the chemicals, monitor their performance, and report the results. Service levels vary from simple testing, monitoring of pumps, and inventory control to sophisticated equipment-oriented chemical addition, control, and reporting programs. Vendors also offer training programs for facilities personnel. The costs of these service programs are usually included in chemical prices, but may, if required by the facility, be stated separately as part of a water treatment contract.

Large manufacturing plants and public utility stations sometimes find that it is cost effective for them to take the service aspects of water treatment in-house. A facility may, for example, contract with a blending company to supply specified formulated water treatment products in bulk. Facility personnel then assume the entire job of feeding chemicals, controlling dosage and system parameters, maintaining equipment, monitoring performance, and making water treatment decisions.

This operation, while often quite successful, makes heavy demands for time and technical knowledge on the part of facility personnel. A consultant may be involved in the design and overall monitoring of the program, but the overall operating burden still falls on the facility. Because of the trend toward plant automation, and the accompanying reductions in required plant personnel, most facilities, and particularly commercial facilities, have elected to leave water treatment service as a shared responsibility of the facility and the vendor. However, rather than leave service decisions to the vendors, larger facilities may now separate costs of products and services, and specify the services to be performed.

#### Full-service water treatment

At the opposite extreme from total in-house water treatment, some facilities contract with a vendor for a "full-service" program. In this mode, the vendor supplies all products and does all the work. Facility personnel take no readings, do no testing, and make no adjustments to the system. In a typical full service program, a vendor service person visits the facility, usually once per month. He or she runs simple water tests, adjusts pumps or adds chemicals manually as needed, replenishes chemical inventory, and prepares a brief report. The vendor may also take quarterly water samples for laboratory analysis and do other routine monitoring. Between these visits, the water systems run unattended unless the vendor is called to respond to a problem.

Full-service contracts appeal mostly to facilities that do not have time or personnel to devote to water treatment operations. These are mostly small commercial buildings and manufacturing plants. The water systems in these facilities are usually straightforward: low-pressure steaming or hot water boilers and simple circulating condenser water and chilled water loops. Correspondingly, basic chemical treatment programs are used and feeding and control systems are elementary. High doses of chemicals are used to help ensure that active levels in the water do not run below range between service visits. Performance monitoring tools such as corrosion coupons may not be used unless specifically requested by the facility.

Some large commercial facilities do write full-service water treatment contracts. In these cases, the vendor may use modern remote monitoring equipment to "watch" the systems from a central office. Also, at least one facility operating engineer will usually be qualified in water treatment and able to recognize and respond to unusual situations.

The major difficulty with full service water treatment is that problems can arise and become serious before they are recognized and corrected. Algae blooms in cooling towers, microbiological fouling and MIC in both condenser water and chilled water systems, and dissolved oxygen corrosion in boilers and hot water systems are examples of problems that can go unnoticed until operations are impacted, unless systems are inspected and tested regularly by trained facility and service personnel.

### Support service water treatment

Support service is by far the most widely used form of vendor water treatment service in medium-to-large facilities. Support service involves joint responsibility for the facility water treatment operations by facility personnel and the vendor. The vendor will supply chemicals and all feeding and monitoring equipment. This equipment can range from simple to highly complex and automated. The vendor will visit the facility at a frequency ranging from weekly to monthly, depending on the needs of the facility and the size and complexity of the water systems. Routine vendor service work is discussed later.

In a support service water treatment program, facility operating engineers are responsible for the daily operation of the water systems. This includes daily chemical testing of the water systems, plus testing each shift of the boiler systems if high-pressure boilers are involved. Daily service work by facility operating engineers is discussed later.

### The role of consultants in contracted water treatment services

Experienced water treatment consultants are assuming an increasingly important role in modern facility water treatment service programs. As chemical programs have become more complex, feeding and control equipment more sophisticated and service programs more elaborate, managers of large facilities have found it advantageous to have a technical person working with and representing them in dealing with the vendor. The consultant serves in several different ways:

1. They prepare specifications and evaluates proposals, as explained later.
2. They help to ensure that chemical programs are appropriate for the facility systems and that all feeding and monitoring equipment is properly installed and maintained.
3. They review facility water treatment logs and vendor service reports, and prepares periodic independent assessments of the status of each system. As part of this assessment, he performs independent water analyses and inspections of facility piping and equipment, and he works cooperatively with the vendor to make necessary corrections.
4. They assist the vendor in recommending needed changes and improvements to water treatment programs and to facility systems or operating procedures.

The role of a water treatment consultant should be as a facilitator working cooperatively with both facilities management and the water treatment vendor, to provide optimum protection for all water systems while at the same time maximizing their performance.

## Installing a New Water Treatment Program

This section discusses the process of selecting a water treatment vendor and installing a new chemical treatment program.

### Selecting a water treatment vendor

Installing a new water treatment program, or replacing an old program, involves as a first step selecting a water treatment vendor. Environmental discharge and air-quality restrictions have narrowed the available choices for water treatment chemicals. As a result, most treatment programs now involve combinations of known corrosion inhibitors, scale and deposit control agents, and microbiocides. The opposite is true, however, of modern chemical feeding and control equipment, and chemical delivery systems. The software associated with operating the systems and collecting and presenting performance data is also unique to each vendor. For these reasons, water treatment vendors are differentiated today more by their equipment options and the quality and quality of their service work than by their chemical programs.

Most facilities select their water treatment vendors based on recommendations and personal contacts, or on sales calls and proposals by the vendors. This process works well, and it has one major advantage. The most important part of any service program is the knowledge, experience, and dedication of the vendor personnel providing the service. Sales calls, interviews, and recommendations provide the opportunity to evaluate the companies and their people as well as the chemical programs and equipment.

Prospective vendors should ask for an opportunity to survey the facility, speak with knowledgeable employees, and study the history of water treatment in the facility, to help them prepare appropriate proposals. The resulting proposals provide insight into each vendor's understanding of the facility and its water treatment needs. It is important to remember that proposals must be individually adapted for each facility. A chemical and service program that works well at one building or manufacturing plant may not be at all suitable for another, similar facility close at hand. Differences in water quality, operating procedures, critical cooling and heating demands, and many other factors can combine to make apparently similar facilities quite different in terms of water treatment needs.

Cost should be only one important factor involved in vendor selection. Vendors will, of course, try to be competitive, but responsible vendors will base their proposals on the products and services they believe are needed for success in the facility. Large differences in the annual costs of proposals should be viewed with suspicion until the proposals have been technically evaluated and found to be sound.

### Water treatment specifications

Larger facilities, and facilities that include critical cooling and heating needs, or complex equipment such as high-pressure boilers, have found that the use of bid specifications and formal requests for proposals (RFP) offers several advantages:

1. It provides a controlled bid environment that puts technical proposals on an equalized basis for evaluation, and may lead to more competitive pricing.
2. It ensures that vendors will have available the critical facts about the facility that affect water treatment, so that proposals may be more realistic.
3. The specification can define specific required services, with vendor options to offer additional services. This helps to limit service proposals to work that is really needed and is cost-effective at the facility.

**Performance specifications.** It is important to remember that water treatment specifications, and the resulting proposals, are different from purely mechanical specifications and proposals. Mechanical projects can be clearly defined in terms of what is needed, what materials are to be used, how the work is to be done, etc. This is not the case with water treatment. In spite of limitations on chemical programs that can be used, there are many options available to the vendors. Each vendor has developed its own product line and application technology, and its own equipment and service programs that it trusts and uses effectively.

Forcing vendors to bid to a specifically defined chemical program makes evaluation of proposals simple. However, it severely limits each vendor's ability to offer the programs the vendor thinks will be most efficient and cost-effective in the facility. It also limits the facility's ability to obtain

**TABLE 9.1 Typical Cooling Water Performance Specification**

| Parameter | Open systems | Closed systems |
|---|---|---|
| Corrosion on mild steel | Less than 2.0 mpy | Less than 1.0 mpy |
| Pitting attack on mild steel | None | None |
| Corrosion on copper alloys | Less than 0.2 mpy | Less than 0.1 mpy |
| Scaling and deposition | None | None |
| Microbiological fouling | No visible deposits | No visible deposits |
| | No health hazards | No health hazards |
| | Total aerobic count $< 1 \times 10^4$/mL | Total aerobic count $< 1 \times 10^3$/mL |

and use the best available technology that meets the facility's needs.

A better way is to set performance specifications and allow each vendor to propose the programs that he or she believes will most cost-effectively meet these specifications. For example, the parameters that determine the performance of facility cooling water treatment programs are: corrosion rates on system metals, mineral scaling, general deposition on surfaces, and microbiological fouling. A typical performance specification for cooling water treatment is shown in Table 9.1.

**Suitability to task.** Performance specifications such as recommended here make the job of evaluating vendor proposals more difficult. The process is further complicated by the fact that there are no standards governing the quality or performance of water treatment products. For example, standards for pipefitting and electrical materials and methods help to ensure that work done according to these standards provides the expected service. That is not the case with water treatment work.

As explained earlier, differences in water quality, facility design, and service requirements make each facility water treatment program unique. Consider, for example, the widely different water treatment needs of a steel mill in Indiana, a power plant in California, a pharmaceutical plant in North Carolina, a financial data center in New York, and a hospital in Texas. All of these facilities use water and need water treatment. However, their requirements differ widely. Specifications and performance requirements written for one of these facilities would have little value for another facility in the group.

Even within narrowly defined groups, such as commercial buildings in Manhattan or power plants on Lake Michigan, for example, water use and performance requirements differ widely. Some buildings simply provide five days per week office space, while others contain financial centers that operate continuously, generate high heat loads, and require precise environmental control. Some power plants use once-through cooling water, while others use recirculating systems with several different configurations.

For all these reasons, vendor proposals must be tailored to the specifications and operating conditions at each facility. Vendors will sometimes propose water treatment programs simply because they are successful elsewhere. Such proposals should be rejected unless the vendor can demonstrate in detail why his or her proposal is appropriate at the facility in question.

By the same logic, chemical feeding methods and control equipment should be selected to match the needs of each facility. For example, a large power station operating with minimum personnel may require the highest possible level of automation in chemical feed and dosage control. At the other extreme, noncritical manufacturing facilities and commercial buildings may not need more than simple controls based on timers and conductivity.

**Facility technical information.** A well-written specification includes the fundamental technical information that all vendors must have in order to prepare their proposals. This can include, for example, the number and type of water systems, volumes of water, special building operating requirements, etc. However, this information will not be sufficient. As explained earlier in this section, each vendor should be expected to survey the facility in order to thoroughly understand the water treatment requirements. Vendor surveys should include equipment inspections, water and deposit analyses, and a review of historical water treatment records, among other work. The vendor's proposal should demonstrate his or her understanding of the facility water systems, current operating problems, and needs for improvement. The proposal should show how the vendor has matched his water treatment program to the specific operating requirements and performance needs of the facility.

**Water treatment contracts.** In addition to full-service water treatment contracts described earlier in this section, support service contracts may be written on a "not-to-exceed" annual basis. The contract price may include both chemicals and service, or the costs of chemicals and service may be specified separately in the contract. The facility may order chemicals as needed, or the vendor may manage the inventory. This can be done either manually or by automated "keep-full" systems as described later. "Not-to-exceed" contracts are usually based on a measure of the facility demand for chemicals, such as annual use of makeup water, number of facility operating days per year, etc. If facility demand cannot be reasonably predicted in this way, the contract may be open ended and written in terms of cost per 1000 gallons of makeup water, per 100 megawatts of power generated, etc.

It is good practice to include required minimum vendor service work in the contract. The vendor may also want to list additional work that the vendor will do, and work that will be done at additional cost as required. This helps to avoid later disagreements over contract provisions and service performance.

## Managing a Facility Water Treatment Program

This section discusses cost-effective management of water treatment service programs, methods for evaluating program performance, and responses to problems that may arise.

### Water treatment service programs

Once a bid has been approved, or a vendor has been selected by another process, a conference should be scheduled to plan the water treatment operation. The purpose of this conference is to plan the service program as a cooperative venture involving both vendor and facility personnel. Serious problems have resulted, for example, from failure of facility personnel to respond to requests from the vendor service representative, or from failure of the vendor to react to information on operating changes supplied by the facility. These are communication errors, which can usually be avoided by developing a team approach to facility water treatment service.

**Service work.** There should be a clear division between the work done by facility operating engineers and vendor representatives. As discussed earlier in this section, the facility is responsible for periodic (usually daily) water testing, data collection from online measurements, adjustments to pumps and control modules, etc. The operating engineers may also remove and replace corrosion coupons, check inventory, place chemical orders, and do other ongoing water treatment work specific to each facility. The vendor provides the necessary training, analytical equipment, and other tools needed for this work. If properly trained, facility operating engineers can also recognize problems, such as cloudy or red water, out-of-range chemical readings, etc., and can alert the vendor.

The vendor representative, during regular service visits, reviews the facility water treatment logs, and the operating logs if appropriate. He or she does independent water testing, including parameters that are not checked regularly by the facility engineers. He or she will calibrate meters, check inventory, and inspect monitoring equipment such as corrosion coupons and heat transfer test units.

During every service visit, the vendor representative discusses water treatment operations with the operating engineers and with facility management, and he or she prepares a written service report. This report contains the results of the vendor's on-site tests, comments on the status of each system, and the vendor's specific recommendations for actions. These can include simple items such as adjustments to chemical feed rates, and more complex issues such as the need to repair or replace equipment, change chemicals or dosages, etc.

The vendor service report becomes the basis for action by the facility operating engineers between service visits. It is critically important that the vendor be objective in his recommendations, and that the facility operating engineers respond quickly to these recommendations. The vendor must never gloss over or try to explain away poor results. Problems must be faced squarely, not to assign responsibility, but to correct the situation before it becomes serious. Similarly, vendor recommendations to repair, replace, or upgrade equipment must be taken seriously. As examples, loss of conductivity control in a cooling tower, or loss of chemical feed to a boiler, can lead to the need for repairs that will far outweigh the cost of the control equipment.

**Documentation.** Documentation is a vital part of the total service program. Documentation begins with the facility water treatment logs and the vendor service reports, discussed above. To these are added performance data: corrosion coupon results, microbiological tests, water and deposit analyses, equipment inspections, and so forth.

The vendor should summarize all of this information in a quarterly report. The quarterly report should also contain a summary of all vendor recommendations, including the actions taken on each recommendation. The quarterly report then becomes the basis for a review meeting with facility management. At this meeting, the status of each system is reported, recommendations are reviewed, and action plans for the next quarter are decided. This meeting helps to ensure that all problems are openly discussed and that well-considered decisions are made. This meeting also provides management with planning information for future repairs, capital improvements, and new water treatment related equipment. Proper use of facility and vendor test data is discussed further later.

In addition to quarterly reports, water treatment vendors may include in their service program an annual report that reviews both technical progress and the actual cost of the water treatment program versus forecasts. These data are useful in contract negotiations and in planning future service programs.

**Training.** As explained above, the water treatment vendor is responsible for training facility operating engineers to perform routine testing and equipment maintenance. The vendor should also provide sufficient general background in water chemistry and equipment operations to help the operating engineers understand the importance of their water treatment work and the need for maintaining careful, accurate records.

In addition to this basic training, vendors may offer specialized courses in the safe handling of equipment and chemicals, pesticide applications, new water treatment technologies, etc. These courses may be offered on-site for large facilities, or off-site in a classroom setting. Facility operating engineers should be encouraged to take advantage of as much training as practical, to improve their understanding of water treatment technology and their ability to recognize and respond to unusual situations as they occur.

### Evaluating water treatment program performance

This section discusses the methods commonly used to evaluate the performance of facility water treatment programs.

**Water and deposit analyses.** Clear, foam-free water with all chemical tests within established control ranges is a reassuring sign that the water systems are in good condition. However, changes in water chemistry are often the first available indication of developing problems.

Facility operating engineers should run periodic water analyses for components that are important in controlling the program. In an open condenser water system, these analyses might include, for example, pH, conductivity, biocide, and chemical treatment level. Boiler water testing might also include silica and hardness. Some of these data may be obtained by automatic on-line analytical equipment and corroborated by periodic tests.

Vendor service representatives repeat these tests and supplement them with other data that indicate system performance. For example, high iron and copper levels are signs of corrosion. Losses in calcium hardness and alkalinity compared to calculated values from the makeup water may be a sign of calcium carbonate or calcium phosphate mineral scaling. Foam may indicate microbiological problems. High levels of turbidity and/or suspended solids should be analyzed to determine the source of the problem.

Facility water treatment log data and vendor service report data should be plotted on trend graphs, such as the condenser water data shown in Fig. 9.2. Some vendors supply software that can plot these graphs directly as data are entered at a keyboard. The graphs clearly show how the test data relate to established ranges. Thus, in Fig. 9.2, the trend in condenser water conductivity compared to the makeup indicates that cycles of concentration are increasing. This could be a normal seasonal variation, or it could indicate a control problem that should be checked. Similarly, the variability in molybdate (chemical treatment) compared to the control range is too wide. Corrosion control and cost performance will both be improved if this variability can be reduced.

In addition to the regular service work, the vendor or a consultant should take quarterly water samples from all systems for complete laboratory analyses. If deposits or sus-

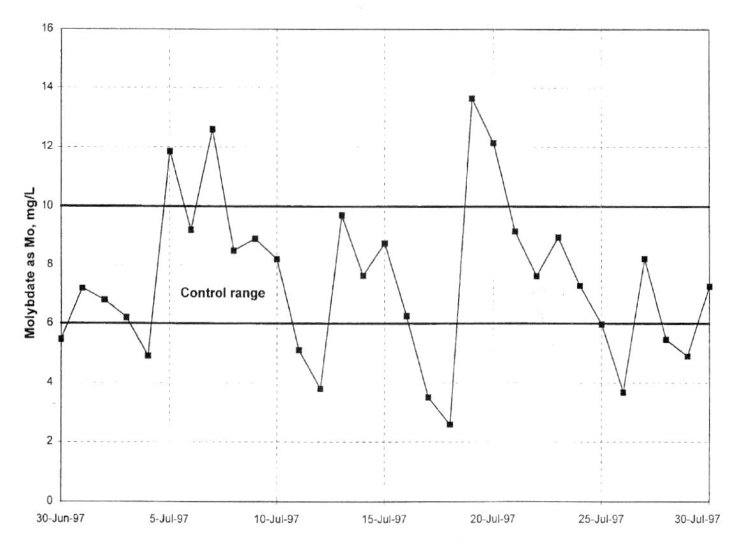


**Figure 9.2** A typical condenser water control chart showing poor control of corrosion inhibitor level.

pended solids are present, these also should be analyzed. For example, Table 9.2 shows typical complete analyses of a makeup water, a condenser water, and a condenser water deposit from a food manufacturing plant. Table 9.2 also shows cycles of concentration calculated by dividing the concentrations of various components in the condenser water by the concentrations in the makeup water. Abnormal or unexpected results from these analyses are highlighted.

Briefly, the data in Table 9.2 show that the condenser water treatment level is good and levels of iron and copper (corrosion products) are low. However, low cycles for calcium and alkalinity compared to nonprecipitating components such as silica and magnesium show that the water is precipitating calcium carbonate. Total microbiological counts in the condenser water are low. However, the deposit analysis shows the composition to be primarily iron oxide, with a very high organic content as indicated by the total organic carbon (TOC) level. This organic carbon could be, in this case, either process contamination or microbiological growth.

**TABLE 9.2 Typical Laboratory Water Analyses of a Makeup Water, an Open Condenser Water, and a Condenser Water Deposit**

| Parameter | As | Makeup Water mg/L | Process Water mg/L | Cycles | Cond.Deposit Wt. % |
|---|---|---|---|---|---|
| pH | pH | 7.46 | 8.25 | | |
| Conductivity | umhos | 87.1 | 730 | 8.4 | |
| M Alkalinity | CaCO3 | 25.7 | 134 | 5.2 | |
| Tot. suspended solids | mg/L | 6.50 | 6.10 | | |
| Aluminum | Al | 0.12 | 0.14 | | 2.84 |
| Calcium | CaCO3 | 60.1 | 295 | 4.9 | 3.51 |
| Copper | Cu | 0.14 | 0.19 | | |
| Iron | Fe | 0.35 | 0.27 | | 63.2 |
| Lead | Pb | <0.01 | <0.01 | | |
| Magnesium | CaCO3 | 8.16 | 64.5 | 7.9 | 1.29 |
| Manganese | Mn | 0.14 | 0.14 | | |
| Molybdenum | Mo | <0.01 | 6.37 | | |
| Potassium | K | 0.58 | 4.80 | 8.3 | |
| Sodium | Na | 7.89 | 75.0 | 9.5 | 3.87 |
| Strontium | Sr | 0.03 | 0.18 | 6.0 | |
| Zinc | Zn | <0.01 | 1.28 | | |
| Bromide | Br | <0.20 | 1.20 | | |
| Chloride | Cl | 30.1 | 240 | 8.0 | |
| Fluoride | F | 0.93 | 7.08 | 7.6 | |
| Nitrate | N | 0.17 | 2.66 | | |
| Nitrite | N | <0.05 | 0.20 | | |
| Orthophosphate | PO4 | 0.05 | 0.21 | | |
| Total phosphate | PO4 | 0.17 | 1.42 | | |
| Silica | SiO2 | 3.60 | 28.4 | 7.9 | 12.6 |
| Sulfate | SO4 | 1.54 | 12.4 | 8.1 | 2.61 |
| Total aerobic count | CFU/ml | | <10,000 | | >10,000,000 |
| Loss on ignition @ 500 F | | | | | 35.4 |
| LSI @ 90 F | LSI | -0.9 | +1.15 | | |
| Cation/anion ratio | | 0.95 | 0.95 | | |

This is one example of how analytical data can be used to detect system problems that would be missed by routine service work. In this case, further work will be needed to determine the source of the iron oxide and the nature of the organic matter, before corrective action can be taken. Early action, however, will help to prevent significant organic fouling and/or microbiological contamination in the system.

**Corrosion monitoring.** Corrosion coupons are by far the most common method for monitoring corrosion rates in facility water systems. Coupons may simply be suspended in the cooling tower basin, but this is not good practice. Aeration and flow conditions in the basin are very different from the piping system, so that coupon corrosion rates measured in the basin are usually not representative of the system.

The best and most common practice is to install corrosion coupons in a rack. Figure 9.3 is a photograph of a typical coupon rack. The rack is made from 1″ diameter mild steel pipe, and has spaces for at least four coupon holders

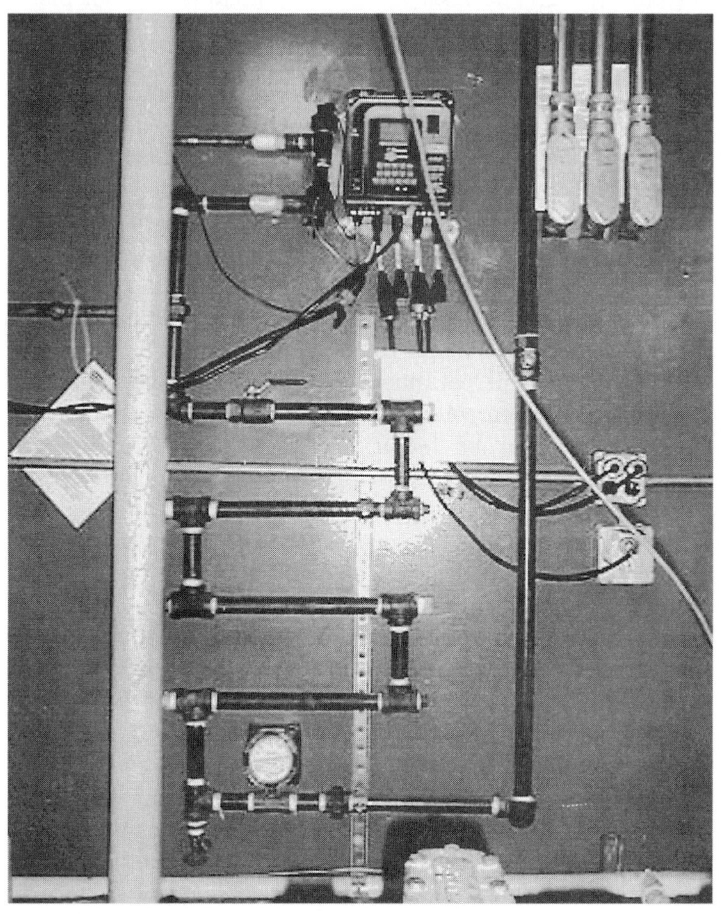

**Figure 9.3** A typical corrosion coupon rack in service.

attached to pipe plugs. The rack must be installed where the pressure drop is sufficient to provide a velocity of at least 3 ft/sec (8 gpm through 1″ pipe).

Corrosion coupons and coupon holder/pipe plug assemblies are normally provided by the water treatment vendor. The vendor may also offer coupon racks fabricated from PVC. These can be used. However, the use of a mild steel coupon rack provides a metal surface, readily available for inspection, that is as close as possible to the condition of the system piping. The rack should also include a spool piece, connected by unions, that can be removed periodically for inspection and then either returned to service or replaced. Finally, the rack should include a flowmeter and appropriate throttling, isolation, and drain valves.

At least one coupon rack should be installed on each condenser water system. In high-rise buildings, this rack should be near the low point of the system. In specific situations, it may also be wise to place coupon racks near the top of the system and on branch lines leading, for example, to tenant space. One rack should be installed on each closed cooling and heating loop.

The vendor will establish a routine for removing and replacing corrosion coupons. This work may be done either by the vendor or by facility personnel. The vendor will process the coupons and provide reports that include both a calculated corrosion rate and a brief description of the condition of the coupon. It is good practice to photograph corrosion coupons as they are removed from the racks. The photographs provide an ongoing record of the condition of each system. As examples, Figs. 9.4 and 9.5 show two sets of corrosion coupons as removed from condenser water racks in high-rise commercial buildings. The coupons in Fig. 9.4 are in excellent condition after exposures ranging from 30 to 90 days, while those in Fig. 9.5 clearly indicate the need for improved water treatment.

Corrosion coupons measure the average corrosion rate on the coupon over the period of exposure. Coupons should be exposed for at least 30 days to allow the metal surface to acclimate to the system. About 120 days comprise a reason-

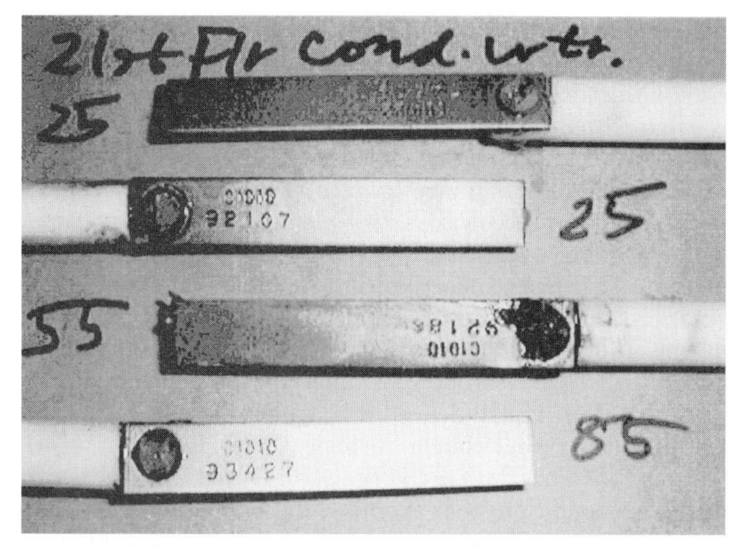


**Figure 9.4** Corrosion coupons exposed in a well-treated condenser water system.

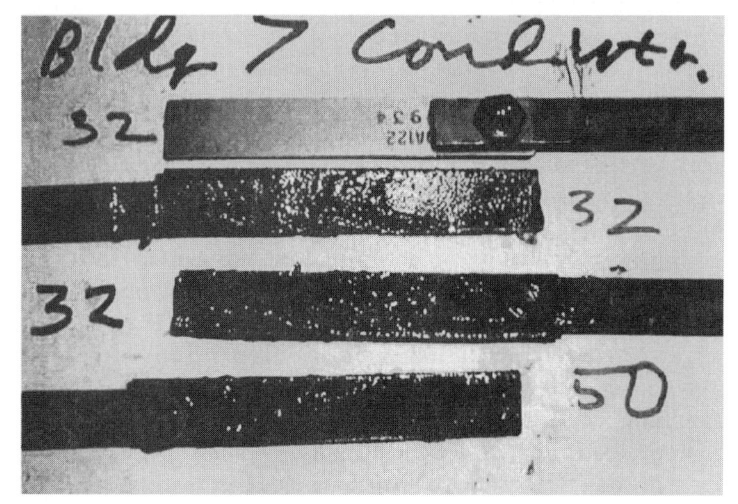


**Figure 9.5** Poor corrosion coupon results from a condenser water system with inadequate water treatment.

able maximum exposure time in condenser water. Longer exposures are acceptable in closed loops that show very low corrosion rates. The vendor will establish a coupon exposure routine to match conditions at the facility.

On-line methods of measuring corrosion rates are also available. Polarization resistance devices apply a small potential difference across two electrodes exposed to the water system, and translate the resulting current flowing between these electrodes into an "instantaneous" corrosion rate. Data may be taken manually or sent to a recorder for a continuous record. The Corratorr® is one commercial example of a polarization resistance device.

Polarization resistance data may not necessarily agree closely with average coupon corrosion rates. However, polarization resistance data are very useful for detecting changes in system conditions. For example, a polarization resistance graph will show an immediate change in corrosion rate due to a change in chemical treatment, long before this change will be recognized in coupon data.

**Mineral scale and deposit monitoring.** Mineral scale is defined as inorganic solids precipitated from the circulating water, as opposed to corrosion products formed by reactions with the metal surfaces. Mineral scales can form in open condenser water systems and in boilers. These systems evaporate water, so that the solubilities of various minerals may be exceeded. Mineral scale rarely forms in closed cooling and heating systems that do not evaporate water. Mineral scales are not often found in areas of the country with very soft, low dissolved solids makeup water. However, in hard water areas, mineral scales may be the major water treatment problem to be handled.

The most common mineral scales are calcium carbonate and calcium phosphate. Both of these minerals are inversely soluble; that is, their solubilities decrease with increasing temperature. For that reason, these scales tend to appear first on heat transfer surfaces, where the water temperature is highest. Depending on the system and the water supply, calcium sulfate, silica, and magnesium silicate can become problems. Magnesium hydroxide can sometimes appear in boilers.

Scaling is controlled by various combinations of makeup water pretreatment to remove scale-forming compounds, and by water treatment additives that help prevent precipitation of specific compounds. These additives, usually low molecular weight organic polymers, also help to prevent solid materials that do precipitate from forming deposits on heat transfer tubes and other surfaces. Water treatment vendors formulate their products with various combinations of proprietary scale inhibitors and dispersants to solve specific problems.

Scale formation in boilers may be recognized by a reduced steaming rate (loss of heat transfer capability) and by periodic boiler inspections. In cooling water systems, on-line monitoring of heat transfer combined with visual inspection of a test heat transfer surface is practical. In a typical deposit monitor, a sidestream from the circulating water is passed across the outside of a heated tube, or through a heated block. This sidestream may be the same water used for the coupon rack. Scale deposits, and also general debris and microbiological deposits, cause changes in heat transfer rate that can be recorded on-line as a "fouling factor." In some devices, the heat transfer surface is visible, so that deposits can be observed as they form.

Figure 9.6 is a photograph of a typical on-line scale and fouling measurement device. Water treatment vendors can supply similar equipment with various levels of instrumentation. Deposit monitors are commonly used in cooling systems where fouling of heat transfer surfaces is expected to be a problem. It is important that the heat transfer surface be representative of the system condenser tubes. If sections of actual condenser tubes are used, then any corrosion that occurs on the heated tubes can also be monitored.

**Microbiological monitoring.** Cooling tower basins are a comfortable place for microorganisms to proliferate. Modern condenser water chemical treatment programs operate in the neutral or alkaline pH range. This pH range and the

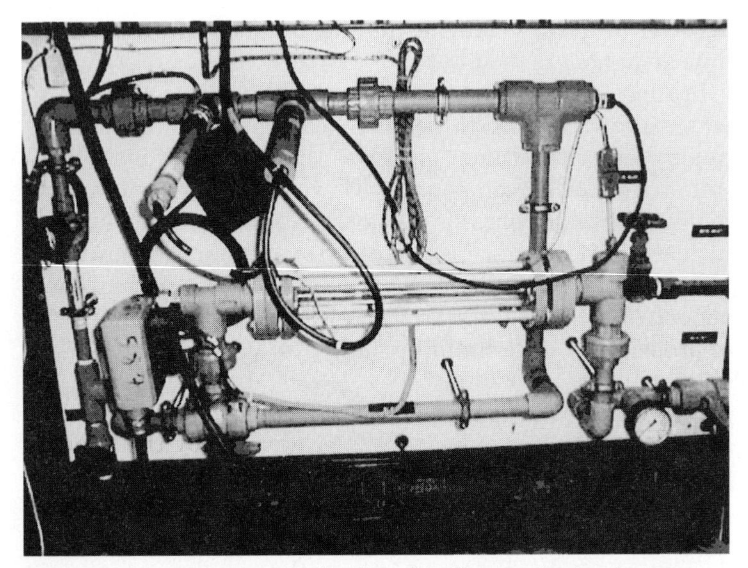

**Figure 9.6** A typical commercial scale and fouling monitoring device.

warm temperatures in the cooling tower basin are ideal for microbiological growth. Chemical treatment programs often contain phosphates, plus various organic compounds that can help to support growth. The air blowing through a cooling tower is a continuing source of both microorganisms and nutrients.

The resulting microbiological slimes that form in cooling towers and on pipes and heat transfer surfaces can interfere with heat transfer and can act as a binder for other solid materials on the surfaces. Serious corrosion can occur beneath these deposits, sometimes made worse by acidic products of microbiological metabolism. This is called *microbiologically influenced corrosion* (MIC).

For these reasons, microbiological control is an essential part of every open cooling water chemical treatment program. Closed cooling systems may also have microbiological problems if the makeup water is contaminated or if process leaks bring organic matter into the system. Two general types of microbiocides are used: oxidizing microbiocides such as sodium hypochlorite that directly attack and destroy cell walls, and various nonoxidizing, metabolic microbiocides that provide slower but sometimes longer lasting kill. The microbiological control program must be compatible with all materials in the system and with other parts of the water treatment program. Also, the complete chemical treatment program must meet all applicable water and air discharge requirements. The water treatment vendor will select specific microbiocides to be used at each facility.

Microbiological monitoring is a critical part of the water treatment service program. It is not practical to sterilize an open cooling tower system. It is, therefore, necessary to maintain microbiological activity at low levels that will not encourage rapid growth and slime formation in the system. Control levels can be set by specification or by the vendor, based on the needs of each system. Monitoring should consist of several stages:

1. Simple "dipstick" testing by facility personnel and the vendor, for total aerobic bacteria.
2. Quarterly laboratory tests of the circulating water and deposits from the cooling tower basin and the system piping, for both aerobic and anaerobic bacteria that could encourage MIC.
3. Periodic testing of the tower water for pathogenic bacteria. The required frequency for this testing will depend on location, opportunities for contamination, and exposure of facility personnel and the public to cooling tower water or to drift from the cooling tower.
4. Inspections of fouling monitors, cooling tower basins and fill, and coupon racks for the presence of microbiological slimes.

**Equipment inspections.** The ultimate test of water treatment program performance, of course, is the condition of the equipment. Cooling towers, chillers, steam condensers, boilers, and other water-using equipment, including the system piping, should be inspected annually. Many different tools are available to aid in these inspections:

1. Inspections of head boxes, tube sheets, boiler drums, etc. provide direct visual evidence of water treatment program performance. Photographs should be taken whenever possible to provide an ongoing record of progress in keeping systems clean and corrosion under control. Vendor photographs should be included in their quarterly technical reports as described earlier.
2. Fiber-optic equipment allows visual inspection of the internal surfaces of condensers, boiler tubes, and pipes. Videotape records can be made of these inspections.
3. Eddy current inspection of condenser and boiler tubes can detect both internal and external damage.
4. Various forms of radiography can be used to check welds and to evaluate the condition of piping beneath insulation.
5. *Ultrasonic thickness* (*UT*) probes can be used to measure pipe wall thicknesses on-line. UT measurements are especially useful in evaluating the condition of old piping systems to determine their general condition.
6. Destructive examination of spool pieces and pipe sections in the laboratory can aid in determining the extent of damage and in predicting the projected remaining life of system equipment. Corrosion coupon rack spool pieces should be destructively examined on a regular basis. System piping should be destructively examined if other testing indicates severe pluggage or metal loss, if failures occur, or as a prerequisite for chemically cleaning a system.

## Cleaning water system equipment and piping

**Defining the problem.** In spite of all best efforts, water systems sometimes become fouled with deposits. Deposits interfere with heat transfer and can restrict flow in piping and heat exchangers. In some cases, water systems that have reached this condition should be cleaned to maintain system performance and to protect the equipment. Also, old systems may require cleaning before a new chemical treatment program can be installed. Deposits that interfere with flow or heat transfer should in most cases be removed. However, thin, hard deposits that protect the surface should be left in place. Removing protective deposits may, in fact, do more harm than good by exposing fresh, unpassivated metal to the system.

The first step in this process is to determine the cleaning method to be used and to estimate the total amount of deposit to be removed. This will require removal of representative pipe sections for destructive examination. This examination will help in determining whether or not sufficient wall thickness remains in the piping system to make cleaning worthwhile. Chemical analysis and physical characteristics of the deposits will aid in selecting the mechanical or chemical cleaning method to be used. Finally, from the amount of deposit in the pipe samples, the total deposit weight density in the system, in ounces per square foot, can be estimated. This will determine the quantity of cleaning chemicals needed.

The following general information on cleaning methods will aid in understanding the processes involved. However, cleaning a facility water system is a site-specific operation. The recommendation to clean should be made by experienced water treatment vendors, cleaning contractors, and consultants who are familiar with the system, with the facility as a whole, and with local safety and environmental regulations.

**On-line chemical cleaning.** Water treatment vendors may recommend cleaning a system on-line, using high concentrations of dispersant chemicals and microbiocides. The microbiocides help to kill and remove biofilms that may be bonding deposits to surfaces, and the dispersants then loosen and disperse these deposits into the water. In condenser water systems, the loosened deposits can settle in the cooling tower basin or they can be removed with the blowdown.

In closed systems, however, some plan must be in place to remove loosened suspended matter that would otherwise settle elsewhere in the system. If the system can be shut down, it can be drained to remove these solids. Several fill and drain cycles may be necessary. An on-line "bleed and feed" process may also be used. This process is slow, wastes chemicals, and may be only partially effective in removing suspended solids. The best method is to use either a permanently installed or temporary sidestream filtration system to help remove suspended solids as they are formed.

On-line dispersant cleaning is less expensive and simpler than any form of off-line cleaning. For that reason, it is often tried as a first step. On-line cleaning is helpful in removing loose deposits, particularly mud and silt, and microbiological deposits. However, this process is not very effective in removing mineral scales or hard corrosion products. Some water treatment vendors also offer stronger cleaning solutions that can be circulated on-line for several days, and then must be drained to remove the cleaning solutions.

**Off-line mechanical cleaning.** In some cooling systems, the main circulating pipes, particularly the vertical risers and large lateral lines, can sometimes be cleaned by a mechanical hydroblasting process. High-pressure water is forced through rotating nozzles that are drawn through the pipes. The resulting slurry of deposits and water is allowed to settle. In most cases, the solids must be carried away for disposal as a toxic waste. The water may be suitable for disposal in the sewers, or, depending on the location, it also may be classified as a toxic waste and require off-site disposal.

Cleaning contractors have the equipment and technology to do this work. Mechanical cleaning is expensive, but it is simpler, faster, and less costly than off-line chemical cleaning. This process will remove all loose deposits, but it will not clean hard mineral scales or tightly adherent corrosion products. Nevertheless, mechanical cleaning may be sufficient. If not, it may at least greatly reduce the amount of deposit that must be chemically removed, thus shortening the time required and reducing the cost of off-line chemical cleaning.

**Off-line chemical cleaning.** The most difficult, expensive, and effective way to remove deposits from a facility water piping system is by off-line chemical cleaning. This is a short-term, high-intensity process. The facility must be evacuated, at least for a weekend. Strong chemicals are brought in, heated and pumped through the system, and then drained and carried away by the contractor. In most cases, the first rinse water must also be disposed of off-site. The entire system being cleaned must be monitored continuously in order to find and immediately clamp any leaks that may develop. The cleaning contractor's personnel will test the cleaning solution frequently and determine when the various stages in the process begin and end.

In spite of these problems and the accompanying high cost, off-line chemical cleaning is often the process of choice for restoring facility water system piping to working condition. The only viable alternative may be replacement of large sections of a piping system. This is especially true in facilities that house data-processing centers and other operations that can only be shut down briefly, with long-term advance planning. The result of a well-done off-line chemical cleaning should be piping that is clean essentially to bare metal and ready for passivation and treatment as a new system.

**Planning the cleaning process.** Any water system cleaning process, from a simple on-line dispersant cleaning to a complete off-line chemical cleaning of a large system, should be carefully thought through, step by step, with a team that includes the cleaning contractor, the water treatment vendor, a consultant, and facility management. Every eventuality must be considered. The result of this planning process should be a flow chart that covers the responsibilities of all parties concerned, from initial facility planning, through arrival of contractor personnel and equipment on-site, to immediate filling and chemical treatment of the system after the final rinse to provide corrosion protection to the freshly cleaned pipes.

The reason for this extraordinary concern is that any unplanned interruption in the cleaning process, particularly an off-line chemical cleaning process, can be very expensive and potentially hazardous. For example, three problems that have occurred during chemical cleaning projects are: (1) sudden appearance of a major pipe failure that cannot be clamped or otherwise repaired; (2) lack of circulation through small diameter lines; and (3) lack of sufficient chemicals on-site to replenish chemicals depleted during the cleaning process. Each of these problems delayed a cleaning project beyond the available window of time, thus forcing the facility to cancel the cleaning project before it was completed. These problems could have been avoided by thorough inspection of the system before cleaning was attempted, and by careful planning and staging of the project.

These cautions are intended, not to discourage chemical cleaning, but to emphasize the importance of careful planning to ensure that the job is done right the first time. Many facility water systems have been successfully chemically cleaned, restoring full operating capability and extending their useful life.

# NOTES

## NOTES

Chapter

# 10

# Architectural, Structural, and Sustaining Maintenance and Repair Services

## Introduction

Effective management of architectural, structural, and sustaining maintenance and repair services can only be achieved through a combination of resources. In architectural, structural, and sustaining maintenance and repair services no level of hard work, expertise, technology, or money is sufficient to accomplish the department's objectives without an effective plan. The plan must integrate elements that make up an interdependent maintenance management system. And it is the facility manager's responsibility to develop and continue to monitor use of this maintenance management system (MMS) to ensure that the changing needs of employees, facilities, and their occupants are addressed.

There are a number of decisions to be made for developing the various types and quantities of elements required in planning and implementing a high-quality MMS. For instance, the elements needed in a MMS for a single-story elementary school are vastly different from those of a high-rise hospital or hotel, just as the MMS elements for a hospital are different from those required of a shopping center. There is no one best system. However, high-quality systems share key elements, and no system can be successful without their use.

## MMS/Work Control

All successful MMSs will have a process for determining the customers and structural and architectural components needs of the facility. The term "customers" refers to all types of facility users. These users may be tenants who are residents, business owners, employees, shoppers, or visitors. The differences between systems for various types of customers are important but the basic elements remain the same. Regardless of the type of customers, the system must include an avenue by which the customers can communicate their needs to the facility department and place this information into meaningful direction for the people who must respond to customer needs. The system must also be able to deal with the facilities' ongoing preventive and predictive maintenance requirements. The more integration of customers in the scheduling and prioritization of these needs, the more successful the system. In responding to requests from customers, the system must be able to differentiate between requests that require an emergency response from requests that can be handled routinely. The political reality is that the system must be established so that the requests are identifiable. While life safety and structural integrity are obvious emergencies, some situations might not be. The system must be clear on emergency responses that require immediate attention. The distinction between emergency and nonemergency responses must be clear for the facility department to know the customers' expectations. The MMS must be organized to meet the needs of different customers. For instance, formal meeting facilities that require impeccable appearance and routine availability could only exist as long as the system is organized so that workers respond immediately to minor problems in these formal areas as though they were emergencies. It is also imperative to know that a "minor" problem in one area may be an emergency in another area for a different function.

The system for collecting information about customer requirements is typically part of a larger organization's "information system." This information system is used not only to collect information and data about customer requirements and preventive maintenance schedules but also to collect information and data related to facility equipment and systems inventory, condition, and maintenance history; occupants needs; and federal, state, and local health and safety rules and regulations. Supporting these systems requires a considerable amount of administrative work whether they are computerized or paper based. However, experience has proven that the value of the information and data available justifies the effort during the decision-making process. Any type of effort to organize information in a MMS must reflect knowledge of the facilities involved. The system must be organized in ways that identify the various facility areas. The system may include an efficient way of identifying areas and room numbers so that workers will know where to perform issued work orders.

## Human Resources

All successfully maintained facilities, regardless of function and customers, have people as the most essential and valuable element of the MMS. Complex computerized maintenance management systems are worthless without qualified people who understand how to use the systems as tools to serve customers. People are needed to monitor sophisticated systems to ensure that they are maintained properly. This element of the MMS benefits customers and the people who deal with customers one-on-one. Remember, it is not the supervisors, managers, or computers who satisfy customers. Front-line employees are the people who satisfy customers.

A number of important things must be considered when determining the knowledge, skills, and abilities that must be represented in the operations and maintenance staff. Technical aspects of the facilities must be taken into consideration so that the skills of the operations and maintenance staff can be matched to needs. While most modern facilities have some computerized systems, they range from simple access control systems to complex, energy management control systems such as those that operate HVAC equipment in response to weather conditions, prearranged schedules, facility use, and utility consumption. In older, less sophisticated facilities, technology may not require expert staff. Yet a facility of modest size and operation may require the expertise of several trades such as carpenters, custodians, electricians, painters, plumbers, and HVAC mechanics. There might be regulations that require jurisdictional credentials/certifications in order to operate and maintain a facility. For instance, facilities with sophisticated HVAC systems may require specially licensed operators. Many single-building facilities can be successfully maintained with just a few people who possess the skill requirements, or good judgment mix, to perform routine work and to access specialists or additional manpower when a job grows increasingly intricate or the volume of work rises. The sophistication of the MMS must parallel the technical maintenance and repair requirements of buildings and facilities.

When there are more than two people working together, leadership must be coordinated. There are basically two ways to organize the human resources involved in managing complex facilities. People can either be organized around their trade and expertise or they can be organized by their specific functions within the facility. Decisions to make about the type of organization are not as important for simpler, smaller facilities as they are for larger, more complex facilities. Traditionally, organizations have been structured by categorizing construction trades where highly skilled supervisors are responsible for planning, assigning, and supervising the work of their specific trade. This type of organization seems to be rooted in specialty trade unions with focuses on work quality control as defined by traditional trade principles. While the quality of work is a critical element in the success of a facilities department, it is not the only criteria by which the organization is judged. Since the mid-1970s there has been a growing interest in creating organizations that are more in tune with facility occupants' needs. In theory, this is being accomplished through what has become known as "zone maintenance organizations." This organizational structure emphasizes the MMS's sensitivity and responsiveness to customers' needs. If a zone system is selected for a facility, a decision must be made to either *decentralize* or *centralize* the operation. One successful example of a decentralized zone used in many facilities includes the establishment of satellite shops. The structure of satellite shops brings workers closer to their customers which leads to shorter response times. A centralized zone system can help to achieve facility department unity and a sense of common purpose and interdependence that are more difficult to achieve in a decentralized organization. These two styles of organization are perhaps extremes. However, neither extreme is right or wrong. Using either approach, the emphasis should still placed on work quality control, communication, accessibility, and customer satisfaction. All are necessary to successful MMS use. Naturally, the system that best accommodates a facility's needs would be the one that strikes the right balance between the two approaches. One important point to recognize is that work quality is not necessarily sacrificed when there is an emphasis on customer satisfaction. Customers will be unhappy if they receive poor quality work. Customers will also be unhappy if the technical quality of the work is good, but those who deliver the work are insensitive to their needs.

Larger facilities will quickly develop the need for additional types of expertise for the work force. Facilities that have lawns and gardens require the support of groundskeepers. And due to the varied tasks of grounds maintenance, there may be a need for varying levels of expertise in this field. Responsibility for grounds maintenance and the equipment used to attain professional results adds a level of complexity to the MMS. Once the facility manager determines the nature of the expertise needed in a field a decision must be made to either hire individuals to perform the required services or outsource the services to a contractor who can produce at a level of expertise.

The decision can be complex when choosing in-house services versus contracting services. Many of the issues previously discussed will be involved in the decision. Decisions concerning delivering service, like all other decisions, will need to be revisited. Service quality is the ultimate criteria involved in deciding whether or not to outsource a service. Getting a service performed by a contractor for less may be attractive. But if the quality of work and customer satisfaction is sacrificed at the expense of saving money, then it is not a good decision.

We will always be required to extract every ounce of value for the dollars we spend. However, we must avoid putting efficiency first and effectiveness second. Doing an effective job efficiently should be the goal of facilities management organizations. Contracting may only involve the number of instances expert contractors needed. If you need to clean windows occasionally, you may want to outsource this service rather than perform this job in-house. Many facility managers have discovered that housekeeping services can be effectively contracted, and that there are a number of very good vendors able to provide a full range of cleaning services. Other contractors are able to provide complete facilities management services at competitive costs in-house. No one option is best for all cases. In fact, the age and condition of the facilities, the customers' expectations, the expertise of facility managers, and the expertise in the local work force must be taken into consideration. If the right information is available, choosing to use in-house or outsource resources is difficult. In fact, if the decision ever appears easy, then some important and relevant piece of information is probably not being considered.

## Customer/Occupant Interaction

No matter the size of the facility, there is a need for both reactive and proactive methods of formal interaction with customers. In order to maintain good relations with customers, a facility department must be able to report facility problems on request, and then follow-up on the status of work in progress. After the work is completed, that department must determine whether or not expectations were met, reporting results to the customer. There should also be a standard and routine form of formal communication between the facility department and customers to ascertain whether or not expectations for services are continually being met. Thus, it is essential for a facility department to constantly adapt its services and improve its quality of work in order to meet the changing requirements of customers.

Being able to understand customer requirements plays an important role in prioritizing and scheduling work requests. Oftentimes those reporting deficiencies do not have sufficient knowledge to provide accurate information about what is wrong. They often do not know or care that a solution to their problem is complex, they just want it fixed. Understanding the functions represented by customers, knowing the nature of customers, and being familiar with the capacities of building equipment and systems are all vital to being able to meet customers' expectations. One of the most difficult realities is that many customers do not know what they expect from workers until workers fail to provide it. An effective facility department needs to have an understanding of its own capabilities so that, through the exchanges involved in routine business, they can help set customer expectations. Customers should be told when to expect a response to their requests, and if possible, who will respond and in what way. One example is responding to a temperature call in the mechanical equipment room. Even if a mechanic goes directly to the equipment and solves the problem, the customers' expectations may not be met until they see and talk to the mechanic who resolved the problem.

## Scheduling

Scheduling repairs and minor alteration work is a critical function that can be facilitated by having available data on the types of activity in the facility. Daily scheduling of minor repair work in a hotel, for example, involves having data on occupied rooms; the time rooms are occupied and checked out; and the time housekeepers clean the rooms. Long-range scheduling must take advantage of the natural cycles of facilities occupants. School year cycles and summer tourist season are other examples that illustrate the information needed to make good scheduling decisions. It is always preferable to involve customers in scheduling decisions, as they know best when their function or operation should be interrupted by repair work. They also know exactly what parts of their operation are being interrupted by lack of repair work. Obviously, real or political emergencies must be handled before routine work. However, facility departments must be careful in balancing emergency and routine requests. Departments that do nothing but respond to emergencies will do nothing more than that, and the number of emergencies will continually rise. There are no facilities without preventive and predictive maintenance requirements. Preventive and predictive maintenance work is important, but may not have the same feel of urgency as an emergency request. Every day of their career, facility managers must fight the battle of urgent versus important. Facility managers are in this business to help people. When they can respond effectively to a request, they feel the accomplishment of helping customers with a problem out of their control.

In the bigger picture, the highest valued activity is implementing the preventive and predictive maintenance programs. These programs prevent the need for the customers' requests in the first place. Preventive and predictive maintenance programs must successfully compete with the urgent and often fun job of responding to customer requests. A "scheduler" in this context refers to anyone who schedules work or job requests. Most people in a facility department in some way play this role. The worker who receives multiple requests must schedule his or her time for each task. Naturally, the front-line supervisor will schedule a higher volume of work as well as more variety.

A department large enough to require a person to perform the primary role of "scheduler" may be involved in scheduling urgent repairs, major remodeling, and preventive and predictive maintenance. Everyone must schedule, and effective scheduling means having access to a wide range of information. The scheduler's task is to constantly strike the right balance between proactive (preventive) and reactive activities. On one hand, too much emphasis on proactivity will result in unhappy customers because customers require an effective response capability, regardless of the long-range effectiveness of proactivity. There will always be the need to respond well. On the other hand, too much emphasis on reactivity will result in the need for more and more reactive attention, as equipment and systems will deteriorate due to lack of scheduled service.

The scheduler must also know the capabilities of the workers performing the work. One worker might be able to handle multiple tasks and make good priority decisions while another might not. Dispatching workers to perform work is another function that requires good information.

Not only does the dispatcher need to know the nature of the work (i.e., whether or not it is critical), they must also know the appropriate worker to dispatch to a particular job. This decision is made more complex because of the marginal quality of information given by customers and the high probability that whoever is selected to respond may be preoccupied on another assignment. Workers, who are in direct contact with customers at the time of the work request, need customer assistance to extract the best possible information about the criticality of the deficiency so that they can communicate this to the scheduler and worker or workers who must respond.

The most frequently asked question by those responding to dispatch is "What emergency do we put off to respond to this call?". The more trades that are involved, the more complex the decision on how to route a work request. Each trade will eventually have to develop a decision tree approach to help those who are managing the workflow determine how best to route a call. Plumbing calls with have a very different decision process than those of locksmiths or electricians. The differences in the expertise of trades persons will also make work routing and dispatch decisions more complex.

Constant feedback is required between workers routing and dispatching work, and those performing the work, to continually improve the information on which the system depends.

Due to the constant flow of requests from facility users and from ongoing preventive and predictive maintenance programs, scheduling of work must be done constantly. At any time of the day a request can be received and may change previously made scheduling decisions. You should not expect to get urgent or important work requests before routine requests. Therefore, a first-in/first-out scheduling model will not be successful in emergency cases. Routine repairs may be handled using the first-in/first-out schedule, but you must always respond to urgent and important work before responding to routine requests. Achieving the right balance between urgent/important response and routine response is important. A routine request that takes too long (from the customer's perspective) may possibly become an urgent or important request, when the customer makes a call to complain. The scheduling system, part of the overall MMS, must utilize existing resources to accomplish all types of work so not to funnel all resources toward the work that is the most urgent and important at the moment.

## Work Documentation

All work must be documented so that the expended flow of resources can be evaluated and understood. Those who have to back-charge for their work need to know the period of time the work was performed, the labor rate, and costs of materials and supplies used. Preventive and predictive maintenance work must be documented so that the effectiveness of the programs can be measured and improvements made as needed. There is also a range of federal, state, and local rules, regulations, and documentation required for personnel, work practices, hazardous substances, and even specific charges for work-related grants or special contracts.

## Safety

In recent years the importance of safety in the work force, and for facility users, has increased and has become the focus of intense federal, state, and local rules and regulations. The level of required training and other safety related program activities is still increasing. Large and more complex facility departments should designate a Safety Programs Coordinator as a collateral duty, while smaller departments might use contractors or even attempt to develop in-house expertise. The increasing changes in requirements make the growth of in-house expertise a very difficult proposition. An example is *internal air quality* (*IAQ*). With the advent of sophisticated HVAC systems came the need for buildings to be sealed essentially from contact with outside air. Technology continues to improve and many of the earlier problems are now being addressed, but there remains a need for the capability of any facility department to respond to IAQ complaints. The response typically involves air testing, HVAC system inspection, and communication with occupants to see how they are using the space, and what types of equipment are being used. As new materials are incorporated into facility components and furniture, the IAQ issue becomes even more important.

### Planning and Estimating

Facility departments must be able to predict the amount of labor and overall costs that are associated with different types of work. The ability to predict and prevent problems means establishing a system that will estimate the costs and the length of time a job will take, and also involves work scheduling and planning. There are several quality estimating systems that can be used to formalize the estimating process. These systems require trained personnel who can integrate the planning and estimating system into the MMS to speed the process and organize the process so that it is easy to use.

### Quality Control/Customer Satisfaction

Finally, facility departments must deal with the issue of quality control and customer satisfaction. The facility manager must know how well he or she is meeting customer expectations. And they must know something about the quality of work being performed by facility workers. The department that relies solely upon unsolicited customer feedback for work quality and customer satisfaction is courting disaster. Problems with work quality, or work not meeting customer expectations, can quickly lead to customer dissatisfaction. The facility will also suffer and low-quality work will cause facility deterioration. It is always best for workers and supervisors in facility departments to have first-hand knowledge about quality and satisfaction problems so that they can be proactive by making changes and communicating with customers rather than waiting for complaints. An effective response to a customer's call can at least control the situation until real improvement has taken place. Focusing continuously on work quality and customer satisfaction has the potential to improve the facilities and the satisfaction of those who use them.

The ultimate goal of any facilities organization is to create an environment which promotes that success and excellence are possible. Whenever facility occupants are hindered or limited by problems in their surroundings, the facilities organization is not meeting their requirements. Successful facility users create opportunities for themselves and for the organizations that interact with them. While the success of your customers might not guarantee your success, it cannot hurt. Nevertheless, not giving customers and facilities proper attention may just guarantee your failure.

# NOTES

# NOTES

Chapter

# 11

# Appraisal of the Facility Department's Performance

## Introduction

Organizations exist to meet a variety of institutional and individual objectives. These objectives are met through a concerted effort by people to convert resources to a product or service, and to implement programs and activities. To ensure steady attainment of objectives, a facility manager must know how well the facility department and its personnel are doing. The process of measurement and evaluation of how shops and individuals are performing is known as *performance appraisal.*

Performance appraisal has received a great deal of attention over the last 30 years. A variety of terms have been used to describe typically the same process. Among those most frequently mentioned are "performance measurement," "performance evaluation," "merit rating," "periodic review," and "periodic assessment." The subject of performance appraisal is one that most often carries negative connotation in organizations. We all enjoy hearing favorable words about our actions, but try to evade those that are not favorable. Honest appraisal, however, carries a little of both.

Performance appraisal is a standard ingredient of all organization life. The question, really, is not whether there will be any performance appraisal, but, rather, what method will be used. Some form of evaluation takes place in every human encounter. We make judgments about other people in terms of how consistent their behavior is with our own or society's beliefs, norms, and values. The same thing happens in organizations. Smaller organizations may utilize a casual, unsystematic approach with little or no documentation, but the process takes place anyway. Larger organizations (and many smaller ones also) tend to use a more formal method of performance appraisal. These organizations have recognized the value of information on the status of their most important resource—people. In addition, employees more actively demand to know where they stand in the organization and how, specifically, they can achieve their career objectives. Pressure from unions to document personnel actions provide a strong boost for the need to have a formal appraisal system.

In order for a machine to carry out its purpose, every single part must perform according to specifications. The same is true with the personnel in the facility department. Therefore, a true appraisal of the facility department must include the assessment of individuals and also the department as a whole.

## Performance Appraisal System

### Introduction

The appraisal of the facility department's performance, as an entity, is more complicated than that of its personnel appraisal. The major reason is that the performance of the facility department is expressed by a large number of attributes, several are objective, the rest are subjective. For example, cost of maintenance and operations, level of service, downtime cost, customers' satisfaction, and employees' satisfaction are all attributes of the overall performance.

Therefore, the appraisal system to be used must be multidimensional in nature.

All methods for measuring an individual's performance are subjective in nature. That is, they express the attitudes or opinions of people. A similar approach can be used in the facility department office and shop's performance evaluation. Through the use of forms or questionnaires, one can collect information and data on performance of the facility department. A most acceptable tool for such a purpose is the *checklist.* Operations, maintenance, and management functions appraisal checklists, presented in Chap. 1, are instruments that pose a series of questions to the facility manager regarding the performance level of his or her facility department. A performance evaluation method that can be used, shown following, is arrayed so that one may choose from among five possible answers for evaluating a specific function topic.

Point values, as shown, can be assigned to the performance assessment of each function topic, and an overall point value can then be computed. Interpretation can be made of the total points accumulated, including excellent, good, fair, or poor performance. Subjective evaluation can be used in conjunction with objective evaluation methods for an overall performance appraisal. Figure 11.1 on pp. 424 to 431 presents sample questions that can be used by the facility manager in evaluating overall department performance arrayed by departmental functions.

**Figure 11.1** Sample performance evaluation questions.

| Department Function and Rated Performance Items Points | Very Poor 0 | Poor 1 | Average 2 | Good 3 | Very Good 4 |
|---|---|---|---|---|---|
| **Outsourcing Functions** | | | | | |
| 1. Outsourcing procedures have clear goals and objectives and objectives | ___ | ___ | ___ | ___ | ___ |
| 2. Functions outsourced include:<br>a. Maintenance functions such as plumbing, mechanical, painting, paving, electrical, grounds, etc.<br>b. Engineering functions such as architectural and/or drawings<br>c. Custodial<br>d. Security<br>e. Trash removal and recycling<br>f. Abatement of safety and/or environmental hazards | ___ | ___ | ___ | ___ | ___ |
| 3. Departmental outsourcing administrator assigned | ___ | ___ | ___ | ___ | ___ |
| 4. Departmental outsourcing alternative methods approach include:<br>a. In-house work force handles day-to-day requirements and contractors handle major peaks<br>b. Minimum in-house work force performs part of day-to-day requirements and contractors handle balance of day-to-day requirements and peak requirements<br>c. No in-house work force assigned. Contractors handle all work requirements | ___ | ___ | ___ | ___ | ___ |

**Figure 11.1** *(Continued)*

| | | | | | |
|---|---|---|---|---|---|
| 5. Contractors are providing good-quality work | ___ | ___ | ___ | ___ | ___ |
| 6. Contractor defined operational procedures are established | ___ | ___ | ___ | ___ | ___ |
| 7. Contractors performing functions now provide good work supervision and technical ability | ___ | ___ | ___ | ___ | ___ |
| 8. Contractor's work has been scheduled and a reporting system has been established that shows the contracted work has been accomplished, per the agreement | ___ | ___ | ___ | ___ | ___ |
| 9. Awarded contracts specify work that is *not* to be accomplished | ___ | ___ | ___ | ___ | ___ |
| **Subtotals** | ___ | ___ | ___ | ___ | ___ |
| **Function's Total** | ___ | | | | |
| **Operations and Maintenance Plans** | | | | | |
| 1. Maintenance management information system (MMIS) developed, implemented, and observed | ___ | ___ | ___ | ___ | ___ |
| 2. Software selected and used based on facility department personnel and their primary mission | ___ | ___ | ___ | ___ | ___ |
| 3. Maintenance service procedures developed, implemented, and observed | ___ | ___ | ___ | ___ | ___ |
| 4. MMIS is integrated with the organization's information system | ___ | ___ | ___ | ___ | ___ |
| 5. Annual Maintenance Operational Report is developed, implemented, and published | ___ | ___ | ___ | ___ | ___ |

*(Continued)*

**Figure 11.1** *(Continued)*

| | | | | | |
|---|---|---|---|---|---|
| 6. Maintenance Management Manual is developed and published | ___ | ___ | ___ | ___ | ___ |
| 7. Management Operational Plan is developed, implemented, and observed | ___ | ___ | ___ | ___ | ___ |
| 8. Building Operational Plan is developed, implemented, and observed | ___ | ___ | ___ | ___ | ___ |
| 9. Comprehensive Facility Operational Plans are developed, implemented, and observed | ___ | ___ | ___ | ___ | ___ |
| 10. Facility Occupant Support Plan is developed, implemented, and observed | ___ | ___ | ___ | ___ | ___ |
| 11. Quality Control Plan is developed, implemented, and observed | ___ | ___ | ___ | ___ | ___ |
| 12. Formally planned and scheduled facility tours are developed, implemented, and observed | ___ | ___ | ___ | ___ | ___ |
| 13. Engineered Performance Standards (EPS), used for work order planning and scheduling, are installed and used | ___ | ___ | ___ | ___ | ___ |
| 14. Indoor Air Quality (IAQ) Plan is developed, implemented, and observed | ___ | ___ | ___ | ___ | ___ |
| **Subtotals** | ___ | ___ | ___ | ___ | ___ |
| **Function's Total** | ___ | | | | |
| **Preventive and Predictive Maintenance Plans** | | | | | |
| 1. Preventive maintenance plan is developed, implemented, and observed | ___ | ___ | ___ | ___ | ___ |

**Figure 11.1** (*Continued*)

| Item | | | | | |
|---|---|---|---|---|---|
| 2. Computerized Maintenance Management System (CMMS) is installed and operating effectively | ___ | ___ | ___ | ___ | ___ |
| 3. Management supports CMMS use | ___ | ___ | ___ | ___ | ___ |
| 4. Personnel are trained in the use of CMMS, willing to use it, and enter *Useful* data | ___ | ___ | ___ | ___ | ___ |
| 5. Equipment and systems cognizant data entered into the CMMS databases | ___ | ___ | ___ | ___ | ___ |
| 6. A computerized facility inventory has been developed | ___ | ___ | ___ | ___ | ___ |
| 7. Facility systems and equipment have been labeled | ___ | ___ | ___ | ___ | ___ |
| 8. Annual and special PM schedules developed, implemented, and observed | ___ | ___ | ___ | ___ | ___ |
| 9. Computerized record keeping and financial and management reporting is used | ___ | ___ | ___ | ___ | ___ |
| 10. Predictive maintenance plan is developed, implemented, and observed | ___ | ___ | ___ | ___ | ___ |
| 11. Latest predictive maintenance technologies are available and used | ___ | ___ | ___ | ___ | ___ |
| 12. Formal lubricant analysis program is developed, implemented, and observed | ___ | ___ | ___ | ___ | ___ |
| **Subtotals** | ___ | ___ | ___ | ___ | ___ |
| **Function's Total** | ___ | | | | |
| **Operations and Maintenance Procedures** | | | | | |
| 1. Initial inventory of building systems, utilities, and controls is developed and keyed into the CMMS databases | ___ | ___ | ___ | ___ | ___ |

(*Continued*)

**Figure 11.1** *(Continued)*

| | | | | | |
|---|---|---|---|---|---|
| 2. Condition assessment, defining needs of installed equipment, has been made | ___ | ___ | ___ | ___ | ___ |
| 3. Design assessment, defining ability of installed equipment to meet current HVAC codes and standards, has been made | ___ | ___ | ___ | ___ | ___ |
| 4. Building Operations Plan (BOP) has been developed and is observed | ___ | ___ | ___ | ___ | ___ |
| 5. Daily equipment start-up and shutdown procedures have been developed and are used | ___ | ___ | ___ | ___ | ___ |
| 6. Operating watch tours procedures have been developed and are used | ___ | ___ | ___ | ___ | ___ |
| 7. Lighting systems used meet aggregate requirements of tenant spaces | ___ | ___ | ___ | ___ | ___ |
| 8. Temperature controls are set to maintain an indoor environment consistent with general industry standards and practices | ___ | ___ | ___ | ___ | ___ |
| 9. Demand control ventilation is used | ___ | ___ | ___ | ___ | ___ |
| 10. Energy Management Control System (EMCS) is used | ___ | ___ | ___ | ___ | ___ |
| 11. Facilities inspection and maintenance program developed, implemented, and observed | ___ | ___ | ___ | ___ | ___ |
| 12. Equipment and systems maintenance and repair procedures developed, implemented, and observed | ___ | ___ | ___ | ___ | ___ |

**Figure 11.1** *(Continued)*

| | | | | | |
|---|---|---|---|---|---|
| 13. Electrical equipment and systems maintenance procedures developed, implemented, and observed | ___ | ___ | ___ | ___ | ___ |
| **Subtotals** | ___ | ___ | ___ | ___ | ___ |
| **Function's Total** | ___ | | | | |
| **Custodial Services** | | | | | |
| 1. Overall general cleaning quality | ___ | ___ | ___ | ___ | ___ |
| 2. List of "specific actions" prepared | ___ | ___ | ___ | ___ | ___ |
| 3. Staffing levels (quantitatively and qualitatively) | ___ | ___ | ___ | ___ | ___ |
| 4. Position descriptions preparation | ___ | ___ | ___ | ___ | ___ |
| 5. Work quality rating program | ___ | ___ | ___ | ___ | ___ |
| 6. Waste-management program execution | ___ | ___ | ___ | ___ | ___ |
| **Subtotals** | ___ | ___ | ___ | ___ | ___ |
| **Function's Total** | ___ | | | | |
| **Contractor Landscaping Services** | | | | | |
| 1. Overall landscape work quality | ___ | ___ | ___ | ___ | ___ |
| 2. Personnel trained in current and accepted horticultural practices | ___ | ___ | ___ | ___ | ___ |
| 3. Compliance with applicable codes and building regulations | ___ | ___ | ___ | ___ | ___ |
| 4. Replacement of existing and new plant material at contractor's expense if caused by negligence or direct act | ___ | ___ | ___ | ___ | ___ |
| 5. Lawn maintenance program developed and executed | ___ | ___ | ___ | ___ | ___ |

*(Continued)*

**Figure 11.1** *(Continued)*

| | | | | | |
|---|---|---|---|---|---|
| 6. Trees, shrubs, and other plantings maintenance program developed and executed | ___ | ___ | ___ | ___ | ___ |
| 7. Pest-management program developed and executed | ___ | ___ | ___ | ___ | ___ |
| 8. Work quality rating program | ___ | ___ | ___ | ___ | ___ |
| **Subtotals** | ___ | ___ | ___ | ___ | ___ |
| **Function's Total** | ___ | | | | |
| **Elevator and Escalator Services** | | | | | |
| 1. Operating signage and instructions for passengers provided as required | ___ | ___ | ___ | ___ | ___ |
| 2. Periodic maintenance provided as required | ___ | ___ | ___ | ___ | ___ |
| 3. Monitor and document equipment performance | ___ | ___ | ___ | ___ | ___ |
| 4. Equipment maintenance program developed and executed | ___ | ___ | ___ | ___ | ___ |
| 5. Contracted maintenance versus in-house maintenance | ___ | ___ | ___ | ___ | ___ |
| 6. Equipment performance monitoring | ___ | ___ | ___ | ___ | ___ |
| 7. Maintenance performance monitoring | ___ | ___ | ___ | ___ | ___ |
| 8. Equipment modernization's need | ___ | ___ | ___ | ___ | ___ |
| 9. Work quality rating program | ___ | ___ | ___ | ___ | ___ |
| **Subtotals** | ___ | ___ | ___ | ___ | ___ |
| **Function's Total** | ___ | | | | |
| **Water Treatment Services** | | | | | |
| 1. Overall water-treatment services quality | ___ | ___ | ___ | ___ | ___ |
| 2. Contracted services versus in-house water treatment | ___ | ___ | ___ | ___ | ___ |

Figure 11.1 *(Continued)*

| | | | | | |
|---|---|---|---|---|---|
| 3. Performance-oriented water treatment specifications | ___ | ___ | ___ | ___ | ___ |
| 4. Evaluating water-treatment program performance | ___ | ___ | ___ | ___ | ___ |
| **Subtotals** | ___ | ___ | ___ | ___ | ___ |
| **Function's Total** | ___ | | | | |

**Architectural, Structural, and Sustaining Maintenance and Repair Services**

| | | | | | |
|---|---|---|---|---|---|
| 1. Process for determining needs of customers and architectural and structural components | ___ | ___ | ___ | ___ | ___ |
| 2. Knowledge, skills, and abilities of technical staff | ___ | ___ | ___ | ___ | ___ |
| 3. Reactive and proactive methods of customer interaction | ___ | ___ | ___ | ___ | ___ |
| 4. Scheduling repair and minor alteration work | ___ | ___ | ___ | ___ | ___ |
| 5. Work documentation | ___ | ___ | ___ | ___ | ___ |
| 6. Safety programs observance | ___ | ___ | ___ | ___ | ___ |
| 7. Work order planning and estimating | ___ | ___ | ___ | ___ | ___ |
| 8. Quality control plan developed, implemented, and observed | ___ | ___ | ___ | ___ | ___ |
| **Subtotals** | ___ | ___ | ___ | ___ | ___ |
| **Function's Total** | ___ | | | | |

# NOTES

# Index

## ABOUT THE AUTHOR

Bernard T. Lewis, an independent Facilities Management Consultant, has a Bachelor of Science in Mechanical/Civil Engineering from the U.S. Military Academy, a Master of Arts in Mathematics from Columbia University, and a Doctorate of Business Administration from Pacific Western University. He is a Registered Professional Engineer and a Certified Plant Engineer. He has been engaged in facilities engineering and management for over 35 years at the U.S. Army, American Machine and Foundry, Western Electric Company, and the U.S. Navy, and on various public and private sector consulting projects. He has published 20 books covering facilities maintenance engineering and management, including this handbook.

## ABOUT THE AUTHOR